EXPLORING ANIMAL CARE CAREERS

EXPLORING ANIMAL CARE CAREERS

By
CHARLOTTE LOBB

RICHARDS ROSEN PRESS, INC.
New York, New York 10010

Published in 1981 by Richards Rosen Press, Inc.
29 East 21st Street, New York, N.Y. 10010

Copyright 1981 by Charlotte Lobb

All rights reserved. No part of this book may be reproduced
in any form without written permission from
the publisher, except by a reviewer.

FIRST EDITION

Library of Congress Cataloging in Publication Data

Lobb, Charlotte.
   Exploring animal care careers.

  (Careers in depth)
  SUMMARY: Examines 20 careers related to animal
care, many of which require little or no post-high
school training.
  1. Animal culture—Vocational guidance.
[1. Animal culture—Vocational guidance.
2. Vocational guidance] I. Title.
SF80.L62   636'.0023   80–17601
ISBN 0–8239–0536–5

Manufactured in the United States of America

Dedicated
to my daughter
Patty
and all other young people
who love animals.

About the Author

CHARLOTTE LOBB is widely recognized as an expert on people and their jobs. She regularly conducts workshops and seminars on job-hunting techniques, résumé writing, assertive interviewing skills, and career goal setting. She has taught in the adult education setting and offers her expertise as a consultant to both industry and nonprofit organizations.

In her community of Torrance, California, Mrs. Lobb is active as a member of the Civil Service Commission. She is the former chairman of the Manpower Advisory Council, the citizen advisory group for the administration of federally funded employment programs. She also has a long-standing interest in the area of unpaid jobs and volunteerism and has written a weekly newspaper column in *The Daily Breeze* that has recruited over 5,000 volunteers for local agencies since 1969. She is also active on the Board of Directors of the South Bay–Harbor Volunteer Bureau.

In addition to the career-guidance books *Exploring Careers Through Volunteerism* (1976), *Exploring Careers Through Part-Time and Summer Employment* (1977), *Exploring Apprenticeship Careers* (1978), and *Exploring Vocational School Careers* (1979), published by Richards Rosen Press, Mrs. Lobb has both fiction and nonfiction magazine articles to her credit and currently writes on assignment for various trade magazines.

Her husband, Charles Lobb, is Assistant Corporate Director of Technical Education for Hughes Aircraft Company. They have two daughters, Carolyn and Patty.

PHOTOS BY THE AUTHOR

Contents

	Introduction	xi
I.	Getting Experience	3

SCIENCE IS IMPORTANT

II.	Animal Health Technician	9
III.	Laboratory Animal Technician	15
IV.	Zoo Animal Keeper	19
V.	Feed Manufacturer	27
VI.	Animal Science Teacher	31

IF YOU LIKE DOGS

VII.	Show Dog Handler	37
VIII.	Groomer	42
IX.	Police Dog Handler	47
X.	Guard Dog Trainer	52
XI.	Guide Dog Trainer	55

LOVE OF HORSES

XII.	Racetrack Trainer and Jockey	63
XIII.	Blacksmith	68
XIV.	Riding Instructor	71
XV.	Pack Station Operator	76

PERFORMING IS THE GAME

XVI.	Hollywood Animal Trainer	83
XVII.	Marine Animal Trainer	88

BUSINESS

| XVIII. | *Pet Store Owner* | 95 |
| XIX. | *Pet Products Salesperson* | 99 |

PUBLIC SERVICE

| XX. | *Animal Shelter Employee* | 103 |

THE ARTS

| XXI. | *Artist/Photographer* | 111 |

Introduction

This book could be subtitled *"What You Can Be If You Can't Be a Veterinarian."* Many of the people I interviewed for the book, successful businessmen, groomers, trainers, and performers, had at one time or another thought about being a veterinarian. Like many who came both before and after them, they found that becoming a veterinarian is a lot harder than they expected.

To become a veterinarian demands first of all a love of animals—an asset all of the people I interviewed had in full measure. It then involves four hard years of heavy science courses at the college level. That is followed by the most difficult task of all—getting into a veterinary school.

In the United States, only about 18.5 percent of people who apply to veterinary schools are admitted. Keep in mind that these are people who already have completed four years of college with, you can bet, very high grades if not straight A records. That translates into 10,000 applicants for veterinary schools each year and more than 8,000 students disappointed.

Although veterinary medicine is an excellent career field with a potential income of easily $50,000 per year, you may also want to consider some alternatives to that goal. And that is what this book is all about—alternatives.

Each of the careers discussed involves animals. To get ideas about what animal care careers are available, I simply brainstormed with myself, checked the yellow pages of the telephone book, and looked for ideas in newspaper and magazine stories. That probably means I have missed some careers that might interest you. But perhaps these twenty-plus ideas will give you something concrete to think about in terms of how your future might look in the animal business.

In the course of interviewing people and observing them at work, I discovered some interesting facts.

People who have chosen to work day in and day out with animals communicate differently with those animals than does the ordinary man in the street. It is a very special kind of communication that might be called love, or respect. The communication is definitely two-way between the human and the animal, whether it is a dolphin, a dog, or an orangutan. Most of the people I talked with indicated that they have always, since childhood, been able to relate to animals in this special way. But, of course, they don't think it is special, since for them it has always been true. I would hope that only those readers of this book who have this special ability with animals would pursue a career in animal care. I know this ability is necessary for success.

Another fact, unfortunately, is that our society does not highly value people in the animal-care business. That means that pay rates tend to be on the low side for many of the jobs. Even the veterinarian earns only a quarter of what a medical doctor is likely to earn. The highest paid people who deal with animals are the very successful racehorse trainers and animal trainers for television and movies. But no one seemed to mind that they were making less money than they might in some other business. Working with animals seems to have its own rewards.

Another indication that society does not particularly value the animal business is that there are 30,000 veterinarians in the United States to handle the medical problems of all the meat animals as well as pets. Each year 30,000 persons graduate from law school to become lawyers.

Something else is true about working with animals to earn a living. It is not like having pets. Under the best of circumstances, you have to discipline your animals, keep them under control. Under the most trying of circumstances, you have to kill your own animals. There is the horse with the broken leg. There is the runt of a litter. There is the sick animal who is in pain. In most jobs that deal with animals, death is a reality.

If you think it is hard, or a bother, to clean up the backyard mess that your dog makes daily, perhaps you had better think of some other career. Everybody, but everybody, in the animal-care business has to "scoop the poop" every day. When you advance you may be able to hire others for this chore, but you can bet that along your career route you will have done your fair share of "scooping."

One of the nice things about animal-care careers is that you can tie in your love of animals with one of your other interests. That is, if you would like to teach as well as work with animals, you can think about a career as a riding instructor (teaching mostly young children), animal science teacher (at the high school or junior high level), or guide dog trainer (usually working with adults).

Below are some ideas about combining your second interest into a career that is just right for you.

Animal Care + Second Interest	=	Your Job
Science/Medicine		Animal Health Technician
		Medical Lab Assistant
		Nutritionist
		Feed Manufacturer
Teaching		Riding Instructor
		Animal Science Teacher
		Guide Dog Trainer
Business		Pet Store Owner
		Pack Station Operator
		Pet Products Salesperson
		Feed Manufacturer
		Kennel Operator
Performing		Marine Animal Trainer
		Hollywood Animal Trainer
		Circus Trainer
Art		Artist/Photographer
		Groomer
Police Work		Police Dog Handler
		Guard Dog Trainer
Mountain Living		Pack Station Operator
Helping People		Guide Dog Trainer

Perhaps you have such a strong interest in a particular type of animal that you would like to specialize. Then you can look at your career selection by considering jobs that relate specifically to that animal.

Animal	Possible Careers
Dogs	Pet Store Owner
Kennel Operator
Groomer
Hollywood Animal Trainer
Guide Dog Trainer
Animal Health Technician
Show Dog Handler
Police Dog Handler
Guard Dog Trainer
Horses	Racehorse Trainer
Jockey
Blacksmith
Riding Instructor
Pack Station Operator
Stable Operator
Animal Health Technician
Exotics	Zoo Animal Keeper
Hollywood Animal Trainer
Fish	Pet Store Owner
Marine Animal Trainer
Aquarium Maintainer

It is also possible to think of animal-care careers in terms of how many years you are willing to go to school after high school.

Schooling	Possible Careers
No additional schooling necessary but training or college may be helpful	Pet Store Owner
Kennel Operator
Animal Trainer
Pack Station Operator
Racehorse Trainer
Jockey
Dog Handler
Pet Products Salesperson

	Animal Photographer/Artist
	Medical Lab Assistant
Two years of college or other specialized training	Zoo Animal Keeper
	Groomer
	Guide Dog Trainer
	Riding Instructor
	Animal Health Technician
	Police Dog Handler
	Guard Dog Trainer
	Blacksmith
Four or more years of college	Feed Manufacturer
	Teacher
	Nutritionist
	Veterinarian

There are some animal-care careers that this book does not cover. I didn't discuss veterinary medicine as a career, because other books do that quite well. I did not discuss the large field of agriculture and raising of meat animals, except for the chapter on feed manufacturing, though clearly it is a large career field. It is more suitable for an entire book unto itself.

What you will find here are ideas. In total numbers, there are not a lot of people who make their living caring for animals. But if you happen to be one of the people who have that special quality and an unsinkable interest in animals, read on. Perhaps you, too, can reach your goal just as the people I have met have achieved theirs.

For those interested in further reading, specifically in the field of veterinary medicine, I recommend any or all of the books by James Herriot including *All Things Bright and Beautiful* and *All Things Wise and Wonderful.* Two books by David Taylor may also be of interest: *Is There a Doctor in the Zoo?* and *Zoo Vet: Adventures of a Wild Animal Doctor.* Girls may find encouragement in the book *No Job for a Lady: The Story of A. Phyllis Lose, VMD,* by Ms. Lose and Daniel Mannix. Your school's career center will have other books about your future as a veterinarian.

EXPLORING ANIMAL CARE CAREERS

CHAPTER I

Getting Experience

When you get to the point of actually applying for a career job in the animal-care business, your prospective boss is going to want to know how much experience you have. He wants to be sure that your desire to work with animals is real, not just a fantasy about how much fun puppies are. Caring for animals in any capacity is serious business. The boss wants to know you take the job seriously, too.

So how do you get started? Time and again as I interviewed people for this book, I found that they had begun their careers as young children. The curator of reptiles at the zoo had a house full of snakes when he was a youngster. It was a hobby that intrigued him from the beginning and eventually became his career. Almost everyone I interviewed had had pets of one sort or another, and you can bet that they took responsibility for those pets. They didn't just let Mom do all the feeding. They did the grooming and cleaned up after the animals. They taught them tricks. And they probably read books about the care of their particular type of pet.

You can begin getting experience at home, too.

Enjoy and work with the pets that your parents permit you to have. Make it a point to train the animal, including obedience training. Inexpensive dog obedience classes are available in most areas through park and recreation departments.

Look around your neighborhood for ways to gain more experience. Let your neighbors know you are available to care for their pets when they go on vacation. You can also offer to bathe the neighborhood dogs. In fact, since many families don't like to bathe their pets, you might be able to start a small grooming business right at home.

Membership in a 4H Club offers a wonderful opportunity to learn about animals and test your interest in the animal-care career field. Through 4H, you can raise sheep, goats, chickens, pigs, cows, or guide dogs. You learn about feeding, exercise, and general care of the animal you select. There are always people around who can answer your questions and help you raise an animal successfully. With farm animals, it even becomes a way to earn spending money.

When you are in high school, take courses that will help you later with your career. If your school offers courses in agriculture or animal husbandry, by all means enroll. Most schools that have these programs have at least a few sheep, a cow, and perhaps a goat. In rural areas the variety and number of animals is larger. You will have an opportunity to learn something about the care of animals, how to shear a sheep, and what to do when an animal becomes sick. You should also take courses in biology and perhaps physiology to help you understand something about what is going on inside animals, including humans.

By the time you are fifteen or sixteen years old, you should be eligible for some volunteer jobs related to animal care. The SPCA (Society for the Prevention of Cruelty to Animals), city- or county-run animal shelters, and privately run shelters often use volunteers. Your tasks might include helping people fill out forms to report a lost animal, reading the want-ad columns for lost animals and trying to match those with found animals in the shelter, or perhaps assisting with the actual care of the animals. You can learn, among other things, to identify breeds of dogs and the laws about animals in your area.

It is also possible to just sort of hang around stables to learn about horses. With any luck at all, the wrangler will put you to work saddling horses, giving them baths, and cleaning the stalls. You could easily work up to leading trail rides.

Some youth organizations such as the Senior Girl Scouts have special badges or awards for those who gain work experience in particular fields. Sometimes veterinarians cooperate with young people who are trying to earn the badge and allow them to help around the clinic, usually doing clean-up chores, answering the phone, making appointments, and holding animals during examination.

By the time you are sixteen (and sometimes earlier), you can begin thinking about part-time paid jobs. Kennels often hire a young person for after-school work to clean the cages and feed the animals. Pet stores may do the same thing, and so may veterinarians with large practices.

A grooming store, too, may need someone for clean-up chores. Feed stores or stores that sell equipment for horses are likely to need some part-time help as sales clerks.

You can turn a part-time job at a snack stand or as a ticket-taker at the zoo or marine animal show into a learning experience. When you are working, you can use your break or lunch times as an opportunity to get to know the animals and the animal keepers. You can ask questions. You can observe.

It really doesn't matter that your volunteer or part-time job involves the least fun part of animal care, the clean-up. There is a lot of that when you work with animals. What does matter is that you are learning if you really like to work around animals all the time or if you are happier just having a nice pet at home. If you find you truly enjoy the business, you will be able to show a future boss that you have been interested in animals for a long time and have taken some action to prepare yourself for your career.

SCIENCE IS IMPORTANT

CHAPTER II

Animal Health Technician

"Loving animals is not enough to make you a good animal health technician," says Janet Young, a certified technician with six months of experience in this career. "You really have to like medicine, too." The future looks very bright for Janet and others who are entering the veterinary medicine field at the paraprofessional level. Over the last ten or more years, the practice of veterinary medicine has become increasingly sophisticated in diagnosis, treatment, and procedures. Veterinarians now in training are learning to utilize the nursing skills of a technician to increase the number of patients they can treat while reducing the stress and strain on themselves. The presence of a technician in the office allows the veterinarian time to focus on the difficult cases and keep up-to-date with research findings and new procedures.

The veterinarian and technician are finding themselves in a working relationship that benefits them both while improving the care of animals.

Janet worked a year in a related job before she made her decision to pursue college studies for certification.

"I had a job at an animal research center," recalls Janet. "I was a caretaker, learning to perform some treatments. I tried working days and going to school at night, but that was too hard. So I quit my job, moved to Los Angeles (to attend Pierce Community College, which offers a two-year Animal Health Technician course), and enrolled full time."

Janet's strong commitment to her career goal helped her get through that first year of study.

"You have to take some introductory courses in chemistry, biology, and microbiology, which are very difficult, and there is lots of competition," explains Janet. "You're in classes with people who are going to be doctors or who will major in chemistry. I worked very hard at studying."

The second year, when the students begin to specialize in animal health courses, may not be easier but it is more practical than the first year. Students gain hands-on experience handling animals and taking samples for laboratory work, and in the final semester they enjoy eight hours a week of actual experience working with a veterinarian

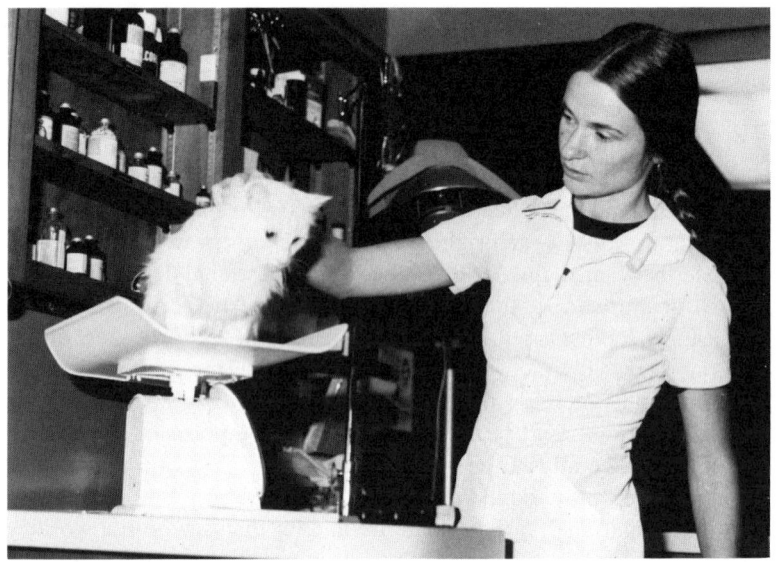

Janet Young, animal health technician, weighs a furry patient.

through an internship program. Students must learn to work with both large (farm) and small animals, although most of the students will spend their careers working in clinics in suburban areas, treating dogs and cats.

In Janet's class of thirty-eight students, only two were men. Male technicians are probably most in demand by veterinarians in large-animal practice, who need help in restraining the animals during treatment.

Janet works at a middle-sized clinic that employs four veterinarians, another AHT who specializes in surgery assistance, three assistants (a less skilled group who work closely with the veterinarians and have typically learned their work on the job), and several clerical employees. The large waiting room is decorated with pictures and sketches of animals. The walls and floors are tile, easy to clean when animals, in their excitement, have an accident. In one corner of the waiting room is a 4′ × 4′ cage containing the clinic's pet cotton-top marmoset, a small monkey. The sign on the cage suggests that the marmoset likes to nibble on any fingers that might be poked through the wire.

At the counter a receptionist greets the clients and directs them to one of four examining rooms. The back doors of these rooms lead to the treatment area where Janet spends most of her time. That area contains cages for her intensive-care patients, examining tables, and all the equipment and medicines she is likely to need in her work except the X-ray machine, which is in a separate room around the corner.

A veterinarian sees the patient first in the examining room. If he or she feels there is a need for lab work or X rays, the animal is referred to Janet, allowing the veterinarian to examine the next patient rather than performing the diagnostic tests himself.

"My work is more varied than if I were a nurse for humans," says Janet. "Most nurses specialize in just one thing. I do blood tests, record vital signs, give injections, insert catheters, give sedations, take X rays and EKG's, and administer fluid therapy or antibiotics. It's a pretty full day if I see twenty to twenty-five patients."

A record of each treatment must be kept. Janet records on the patient's chart how much and what kind of medicine she administers plus the results of any tests. She is also responsible, though everyone is involved to some degree, for maintaining inventory records and knowing when to reorder supplies.

Janet also wears gloves and a face mask to perform autopsies when necessary.

"First of all, you have to know what normal tissues look like," explains Janet, "and then look for any abnormalities. You learn to talk about the tissues in a whole different language of medicine. Then that information helps a pathologist explain the cause of death. I'm interested in physiology. I really want to know why an abnormality occurs, what it means. If I didn't think of it that way, an autopsy might be hard to do."

Janet also does a little animal dentistry, usually cleaning dogs' teeth using ultrasonic equipment. For more serious problems, like the case of a sixteen-year-old macaque (a monkey) with an abcess, the veterinarian performs necessary tooth extractions or calls in a dentist for consultation.

Patients returning after treatment or surgery often see Janet rather than the veterinarian for such procedures as removing stitches, checking for infections, or changing bandages.

Blood transfusions are a frequent part of the job. For animals such as cats, fresh whole blood, not plasma, provides the best results. The clinic has a list of about fifteen cats who live in the area whose owners will permit them to give blood. It's a little like a Red Cross Blood Bank on four legs.

Blood transfusions can become quite a problem when an exotic animal needs help. For instance, the clinic couldn't save the life of a $3,200 tropical bird that was anemic because there isn't a ready supply of blood for this very rare bird.

One of Janet's most important responsibilities is to talk with the owners of the pets that are being treated. Janet has the time to help the pet owner understand the causes of the problem, what the treatment may be, and what part the owner will play in the treatment. Because of her calm, reassuring personality, Janet also provides comfort for clients whose pet is seriously ill.

Janet particularly likes the fact that the veterinarians in her clinic treat exotic animals. They treat a good many birds, which are very delicate to work on. Snakes are common patients, and they see an occasional monkey or raccoon. It all adds to the variety of the job.

Of course, some of Janet's work is routine and not all of it is pleasant.

"Everyone is responsible for a dirty cage," explains Janet. "If you walk by a cage with a mess in it, particularly if there is an animal in the cage, you are supposed to clean it up."

Performing an autopsy can be very smelly work, and there are risks when dealing with either live or dead animals. It is possible to catch certain animal diseases, and all animals have a tendency to bite or scratch if they are hurt or frightened.

Sometimes Janet's work takes her into other parts of the clinic.

There is a special quarantine room for animals with contagious diseases. The room has its own ventilation system so that bacteria are not drawn out into the rest of the clinic to infect other animals.

There is also an operating room. It looks very much like an operating room in a human hospital, with sterilization equipment, overhead lights, and two operating tables. The tables are different, however, since a veterinarian's patients can't lie flat on their backs for abdominal surgery. These tables form a V shape to hold the animal in the proper position. Janet's learning will probably never stop. Even now, one evening a week the clinic's staff meets with an expert radiologist to learn more about the reading of X rays. Soon Janet will be able to help interpret X rays herself, and eventually she can become a licensed X-ray technician.

As more and more veterinarians learn to use animal health technicians in their practice, the demand for technicians should increase and the salary level should also move up somewhat. Currently, a good starting salary for a new graduate of a two-year community college program would be $800 per month. A graduate of a six-month private vocational school course would probably be offered less.

"This is such a new field that you can almost make of it what you want to," says Janet. "By searching around a little, you can find a veterinarian who will use your skills effectively. I find it is a very satisfying career."

Both veterinarians and technicians compare working with animals to working with people. They often conclude that animals are the better species.

SCHOOLS THAT OFFER ANIMAL HEALTH TECHNICIAN TRAINING

Based on somewhat limited information, the following schools offer training to become an animal health technician. Because courses do change, you should request a current catalog from the school's admissions office.

Los Angeles Pierce College
6201 Winnetka Avenue
Woodland Hills, California 91364

San Diego City College
1425 Russ Boulevard
San Diego, California 92101

University of Maine
Winslow Hall
Orono, Maine 04473

Stockbridge School of Agriculture
University of Massachusetts
Amherst, Massachusetts 01002

Institute of Agricultural Technology
Michigan State University
East Lansing, Michigan 48823

University of Nebraska
School of Technical Agriculture
Curtis, Nebraska 69025

New Mexico State University
Agricultural Institute
P.O. Box 3501
University Park Branch
Las Cruces, New Mexico 88001

State University of New York
Agriculture and Technical College at Delhi
Delhi, New York 13753

Central Carolina Technical Institute
P.O. Box 27
Sanford, North Carolina 27330

CHAPTER III

Laboratory Animal Technician

Over the next few months research on 150 pregnant sheep may provide answers to how birth defects happen and how to prevent those defects. Experimental surgery performed on dogs will lead to longer lives for human heart patients. Research conducted on white rats will answer questions about how to control the spread of cancer in humans.

Medical research using animals saves human lives and improves the quality of life. The polio vaccines were all tested first on animals. Animals, not people, were the first to receive heart transplants.

Wherever medical research is conducted, there must be animals and laboratory animal technicians to help in that important work. This includes major hospitals where doctors receive their training and pharmaceutical companies that develop new drugs.

"Actually, there are more people now doing medical research than ten or fifteen years ago," says Keith McDaniel, director of the animal care facility at a major medical research center and president of the Association of Laboratory Animal Science, "but fewer animals are being used. The investigators use other techniques, such as computer studies, as much as possible."

Keith has been in the animal-care business for a long time. He served for twenty-two years in the Navy as a technician involved in medical research. After retirement from the military, he found similar work in civilian hospitals.

"I have a certain empathy for animals, and I'm concerned about the welfare of animals," explains Keith. "Most animals used in research are treated more humanely than those in shelters or even in homes. We don't, for instance, let our animals run loose where they might be run over by a car. We utilize them in a very humane way. I require the same level of care for the animals as is given human patients in the hospital."

It is expensive to use animals in research. A white rat may cost up to $18 from an approved vendor who can provide a disease-free animal. It's not like running down to the nearest pet store to pick up a pet rat for $1.50.

Animal welfare laws, written to insure that animals are humanely treated, set many restrictions on researchers that help protect the animals and also raise costs. As an example, a mouse must be provided with at least 15 square inches of floor space and 5 inches of height in its cage. Dogs, depending upon their size, must be allowed 8 to 20 square feet of floor space and have enough room to stand erect. In Keith's operation, most of the animals have more room than these minimum standards require.

The animals must have food and water available at all times, and cages or enclosures must be kept clean seven days a week. That is the job of the laboratory animal technicians.

Activity begins in the animal rooms between 5:00 and 6:00 A.M., as technicians arrive and change from street clothes to work overalls. First the technicians check to see if any animal has died during the night. If so, the animal is removed from the cage, tagged, and placed in a container that is sent to the morgue for a later autopsy so that the researcher will have information about the cause of death. The technician is also responsible for the environment—the temperature and humidity—in the cage rooms. Then the technician begins the cage cleaning process and restocks the food and water for each animal.

Cages for rabbits and white rats are stacked in rows six feet high. Beneath each wire cage, a tray catches the feces and is cleaned on a regular schedule. Twice a week, the cages themselves are sanitized in an oversized dishwasher.

The process is similar in the dog area, except that everything is much larger and the cages are not stacked. The adult dogs, purchased from shelters when unclaimed after seven days, are in enclosures in pairs. Most of them former household pets, the dogs are eager and excited when people come around to visit. They bark and wag their tails, setting up quite a roar in the barn-sized room. None of the dogs cower in the back of their cages as if they had been mistreated, even after they have had surgery or other treatments.

After the feeding and cleaning procedures are completed, the technician may receive new animals, marking them with tattoos or ear tags and placing an identifying card on the outside of the cage.

Although many of the technician's tasks are routine, "he must be custodian, friend, nurse, and policeman for these animals," says Keith. "And he becomes the eyes of the medical investigator. The researcher may only spend a half hour in the lab each day. It is the technician who has to notice when an animal has lost its appetite or is lethargic and communicate that information to the researcher."

In spite of the routine nature of cleaning and feeding animals, some surprises do happen.

"A monkey may spend months working a bolt loose on a cage and let himself out," says Keith. "Or a dog may become an escape artist and climb out of an enclosure. Worse yet, a researcher may do some work during the night and mix up one batch of rats with another, even ruining someone else's experiment."

There is also some danger when working with animals. There is a risk of catching a disease such as Monkey B virus, which doesn't bother the monkey but is fatal to humans. A bite from any animal may cause tetanus.

"You have to remember that any animal has the potential to inflict harm," explains Keith.

While Keith McDaniel manages the overall operation of the laboratory animal section, Gene Policastri supervises the day-to-day activities in the laboratories.

"This work is the most interesting when we have many research projects going and we have to adapt cages for special problems," says Gene, who has a college degree in large-animal husbandry and supervises fourteen caretaker technicians. "When there isn't much going on, it's hard to get any job satisfaction."

Both Keith and Gene encourage their employees to upgrade their skills to become certified technicians. The Association of Laboratory Animal Science offers three levels of certification: assistant laboratory animal technician, laboratory animal technician, and laboratory animal technologist. Through on-the-job training, home study, and special courses offered to those in the field, trainees learn about care and feeding requirements for typical laboratory animals, breeding and gestation periods, litter sizes, and the effects of stress on animals.

Keith is an instructor in one of these special courses.

About half of his twenty-eight students have a college degree. The others have little or no education beyond the high school level. In time, all technicians working with animals will be required to be certified at some level of competency, according to Keith and Gene.

"Not everyone can work with animals or in research," says Keith. "The work is dirty and smelly. It is often hard for the employee to identify shoveling feces with the aims and rewards of medical research. Many animals do remain in research for years and die a natural death, but for most, death is the by-product of research. Sometimes a technician develops a personal relationship with a long-term animal, and when that animal must be sacrificed (killed to determine the results of the experiment), it takes something out of the person emotionally."

Starting pay rates for animal technicians are relatively low, and opportunities for advancement are somewhat limited.

Hospitals or pharmaceutical companies hire persons with no experience to perform the technician chores at about $650 per month. A senior technician who is certified can earn about $1,000 per month, and a principal technician at a university research center may earn $1,500 per month.

Keith suggests that people who are interested in their first job in this career simply apply to the personnel departments of major medical facilities. It would be even better to talk with people like Keith, to ask questions about the work and express an interest in working in the field by leaving a brief résumé of your background and experience with animals with the boss who has the power to hire.

CHAPTER IV

Zoo Animal Keeper

Spitting orangutans. Poisonous snakes. Exotic birds. Motherless wolves. Any of these may be the daily companions of the animal keepers at the zoo.

All of the keepers, no matter where they work, have one thing in common: they like animals and all the things animals do. They have long since passed the stage of seeing an animal and thinking to themselves, "Isn't hc cute!"

Animal keepers feed the animals and clean their cages and public display areas, but even more important, they observe the animals closely. It is the animal keeper who is first to notice a sick animal. Wild animals don't tend to show sickness. In the wild, a sign of weakness or illness means death. So animal keepers learn to look for very subtle signs—a little less gleam in the eye, sitting in a different place in the display area, loss of appetite, or temper tantrums from a normally calm animal. Early recognition of illness means that the zoo veterinarian has a chance to cure the animal before death occurs or a disease is spread.

Bob Barnes is a senior animal keeper at the Los Angeles Zoo. He has been in the business for ten years and has a master's degree in biology. As a senior keeper, his job is to supervise ten animal keepers who work in the children's petting zoo and nursery. Bob's specialty is mammals.

"Mammals are a little closer to humans than are reptiles or birds," Bob points out, "and therefore easier to understand. When you start dealing with mammals that carry their young in a pouch or egg-laying mammals, then you're talking about a very primitive animal that is much harder to understand. Apes are the most intelligent animals I deal with, though a lot of people feel dolphins are even more intelligent."

When Bob comes to work at 8:30 each morning, he checks his section to find out if anything unusual has happened during the night. A day can start in an exciting way, as it did once when the keepers discovered that the zebras had broken the lock on their enclosure and gone for an evening stroll. Then it was round-up time, with all the keepers playing the role of African cowboys.

Bob's check-in with the nursery is usually a pleasant stop.

"Probably the most rewarding part of the job is seeing a baby animal

Clean-up is an ever-present problem at a zoo.

raised and returned to the big zoo with a minimum of problems," says Bob.

Animals may be raised in the nursery rather than with their mothers for several reasons. Some animals living in captivity don't know how to mother their young and may reject them or even kill them. In other cases, a litter may be too large for a mother to handle successfully. Or zoo management may decide that some animals will be hand-raised in the nursery so that they can be sold or traded to other zoos or to animal trainers who might use them in television commercials. (Los Angeles Zoo rarely sells animals for commercial purposes.) The money then can be used to purchase other rare animals for display.

"You never know in the nursery when the phone will ring and you'll be told to warm up an incubator for some animal who needs help," says Birdie Foster, who has been a part-time animal keeper for ten years.

Like all the keepers, Birdie's job is to feed and clean up after her animals. But with baby animals there are some differences.

"Taking care of a young ape is very much like taking care of a human child," says Birdie. "As infants, their needs are identical. As they grow older, they respond to their surroundings like a preschool child. Fortunately, I don't have to teach them manners. We have only two rules. They are scolded if they try to bite the keeper, spanked on the nose, and they may not leave the cage when the door is open. We try to be very consistent with these rules."

According to Bob, smart animals such as orangutans can actually be spoiled in the nursery setting. "They whine and cry for attention. If that works, it reinforces the behavior and they end up spoiled."

One of the problems of raising animals in the nursery is to convince the baby that it is an animal, not a human.

"We try to place them in cages with animals of their own or similar kind," says Birdie. "When we introduce them to a new animal, we have to be very reassuring so they won't be frightened."

Other animals, like those of the canine family, are less difficult than primates. "They tend to get snappy at the keepers' hands and you have to break up family squabbles, but they don't need the same kind of tender loving care primates do," says Birdie.

Returning a young animal to its family in the main zoo is almost more art than science. Rejection is always a possibility.

"It's a contest between our judgment and the animal's behavior to know when the time is right for a gorilla or chimpanzee," says Bob. "An orangutan will usually stay in the nursery for two to two and a half years, or until he decides to break the glass to let himself out."

Introductions to the new family are done slowly and carefully. First the youngster is placed in visual contact with the adults but safe from physical harm. If this goes well, the animal joins the group in the display area, watched carefully by the keepers. The results are hard to predict.

"We introduced a female marbled cat from the nursery to the male on exhibit," recalls Bob. "I was sure it would take some time for the adjustment. I walked around front and couldn't see either one of them.

Then I realized that they were both inside the same hollow log, just like old friends."

Bob, along with most of the other keepers, is proud of the record of the Los Angeles Zoo in births, particularly among the endangered species. In order to mate successfully, animals in captivity have to overcome, with the help of their keepers, problems of compatibility, having just the right nutritional balance and stimuli, and often the lack of mating and rearing example from their elders. To encourage

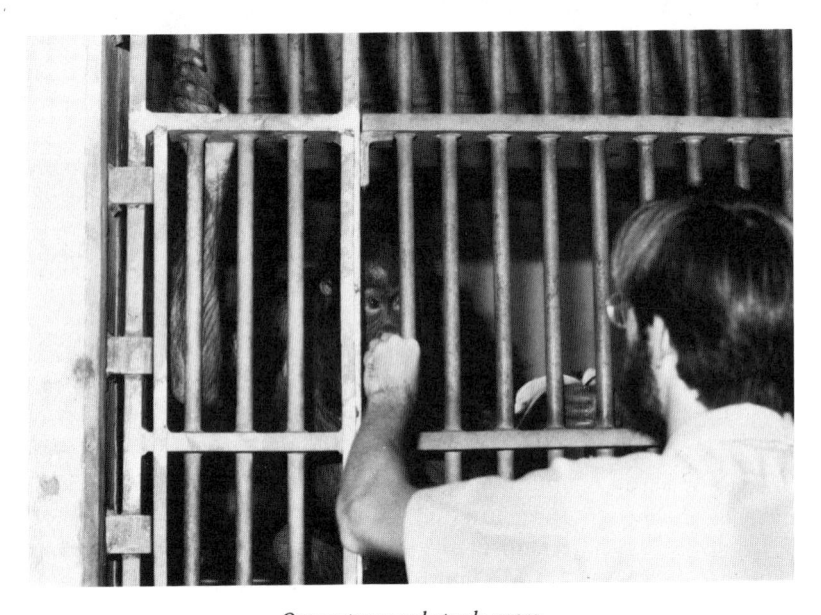

Orangutans need sturdy cages.

mating, the keepers may have to redesign the exhibit to provide a different type of nesting area or change the diet.

"It's good to see babies born," concludes Bob.

Bob often visits the main zoo to check on the graduates of the nursery or just to keep up with the adult animals.

On one particular day Jim Cook was the keeper in charge of the string of animals that included the orangutans and the bears. Jim has been a keeper for two years. He has a college degree in biology. He is assigned as a relief keeper, which means he works several different strings of animals when the regular keeper has his or her days off.

Each morning Jim's first job is to clean the exhibit area, that part of the zoo the public sees, while the animals are in their night cages out of public view. With these potentially dangerous animals, Jim doesn't want to be in the exhibit area while the animals are present.

"I have only once had to go out into the exhibit when the orangs were out, to bring in a macaque that was stillborn," recalls Jim. "One keeper distracted the orangutan with a hose, squirting water on her face so she couldn't see well. I went out in a hurry and helped carry back the dead animal. I wouldn't want to do that again."

After cleaning the exhibit area, Jim moves behind the scenes to release the animals from their night cages. The orangutan cages are sturdy, the gray bars a good inch in diameter. There are double locks on almost everything.

For a special treat and to encourage the orangs to have some physical activity in the exhibit area, Jim has Stacy Gordon, a seventeen-year-old volunteer keeper who is helping him, cut up some apples and distribute them around the exhibit before he releases them. "It encourages the orangs to leave their cages and makes them act like orangs for at least a little while," says Jim.

Next Jim distributes vitamins and medication, as needed, to his charges.

A pill-taking orangutan is a sight to see. Sally, an adult female, balances her two pills on her huge protruding lower lip. She seems to test or maybe taste them with her upper lip. Found acceptable, the pills vanish as Sally's lower lip folds into her oversized mouth.

Now Jim raises the sturdy doors to permit Sally and her friends to leave the cages for the exhibit area. Sally doesn't want to go. One orang leaves and then comes back to see what the delay is all about. Jim resorts to a squirt from the water hose to encourage the departure of the orangs.

"With these smarter animals, the keeper can actually develop some rapport, or at least work better with one than another," says Jim. "The regular keeper, who happens to be a woman, gets along fine with all the female orangs, but the males give her trouble. For me, the males are no problem but the females are sometimes difficult."

Jim doesn't think this is a sexist attitude on the part of the orangutans but rather just a question of personalities. Sally, for instance, is the same orang who once got hold of a rake Jim had left leaning against a nearby wall in her cage area.

"It was out of her reach," recalls Jim, "but she used a towel to whip it around and drag it to her. She got the rake inside her cage and proceeded to methodically break one overhead light after another. She was screaming. I was screaming. It was quite a scene."

After checking to see that the orangutans are all safely in the exhibit area, it is time for Jim to clean the cages.

"In the wild, the orangs live in trees and never see any mess they make," explains Jim. "That's not possible in a zoo."

Jim puts on the black knee-high rubber boots he wears when he hoses down the cages. It's not one of his more pleasant duties and it does smell. And he has just discovered a hole in his right boot.

After cleaning all the cages in his charge, Jim supervises Stacy, the volunteer, as she prepares the evening meals.

Each morning, appropriate types and quantities of food are delivered by the central commissary (kitchen) to the animal areas. For a family of ring-tailed coatis, Stacy finds a package of raw ground meat, perhaps three pounds, a package of fresh fruit including grapes and apples, some bananas, which she peels, and a half dozen hard-cooked eggs. The coatis eat only the yolks of the eggs, so Stacy peels off the whites and discards them, saving Jim a mess he would have had to clean up the next day.

Diets for the animals are carefully controlled. Both fresh food and dried products may be used, depending upon the needs of the animal. Some of the food looks like dried dog food that you might feed to your pet at home.

During the early afternoon, Jim uses his time to take care of any special problems his animals may have or to discuss the care of the animals with senior keepers or curators.

About 4:30 or 5:00 P.M., Jim begins letting his animals back into their night cages. Since the animals have learned that food is waiting for them inside the cage, they usually are happy to return. And for most of them, the cage means security and home.

The potential for danger seems to be one of the things that keep the keepers interested in their jobs and certainly keep them alert.

"You can never take an animal for granted," points out senior keeper Bob. "Some animals lose their tameness as they get older. You can't be careless about closing doors, because these animals are more than willing to go through them."

Most of the potentially dangerous situations seem to occur when keepers must transfer an animal from one area to another. Fortunately, most of the incidents result in eventual laughter rather than tragedy.

One keeper was carrying Eloise, a small female orangutan, from the infirmary back to her exhibit. For some reason, Eloise became frightened and nervous. With her nimble fingers she unbuckled the keeper's belt and lowered his pants while clinging so tightly to the keeper that he couldn't defend himself. The woman veterinarian who was supervising the transfer was laughing so hard that she was of no help at all to the unfortunate keeper.

A similar incident occurred with a woman keeper who was warmly hugged and had her pants lowered by a young bear cub. An off-duty senior keeper came along just in time to save her from public embarrassment.

Becoming an animal keeper at a zoo is not easy. The competition is keen, largely because there are only 450 animal keeper and animal keeper trainee jobs available annually in the whole United States, as estimated by the American Association of Zoological Parks and Aquariums.

A college degree is not required to become an animal keeper, but it is certainly helpful. Officially in Los Angeles, animal keepers are required to have either six months of experience working with exotic animals or satisfactory completion of a zoology course given at the zoo about every two years just before the testing of applicants for the job. According to Bob, there are several ways to obtain work experience.

"It might be work in a pet store or for a veterinarian who handles more than just dogs and cats," says Bob. He also points out that people can, at least at the Los Angeles Zoo, be volunteer animal keepers and learn by doing.

Many of the keepers have gained practical experience on their own. One man worked as an elephant keeper for a circus before moving to zoo work. Another young man had raised wolves and snakes and spent a great deal of time observing animals in the wild.

According to Bob, one book well worth studying for those interested in this career is *Management of Wild Mammals in Captivity,* published by The University of Chicago Press, 1964.

A few community colleges now offer specific courses for potential zoo animal keepers or technicians.

Bob believes that animal keepers are becoming more and more professional, and, although employers are seeking people with practical experience, academic background in biology or zoology is very helpful. He encourages anyone interested in the field to attend college.

Animal keepers are not likely to get rich. Salaries range from $5,000 to $18,000 annually. Not surprisingly, large zoos, particularly in the Northeast and West, tend to pay the most.

Animal keepers are eligible for promotion to senior keeper and eventually to curator or even director of the zoo. At each level a college background becomes more important.

Although there may be debate as to the morality of capturing and caging wild animals, it is clear that for those animals in danger of extinction zoos are the last hope. Human activity around the world continues to encroach on the natural habitat of animals, causing the extinction of mammal species at an ever-increasing rate. An average of one mammal species has become extinct each twelve months since 1900.

Zoos and zoo keepers, through care and breeding programs, enable some of these species to survive.

Psychology students and their professor at San Diego State University have become involved in the care of zoo animals.

"I was concerned that these animals were going slowly stark, raving mad from boredom, and I wanted zoo visitors to see what animals can do, not just what they look like," says the professor. His answer to the problem was to develop toys, starting first with monkeys and apes. To keep the animal's mind active, special feeding tubes were developed that shoot out a biscuit if the animal pushes a certain lever. For toys, he needs durable materials that can take ape-handling such as conveyor-belt rubber and used tuna-fishing nets. The heavy nets of two-inch mesh have been stretched tightly to bounce and slide on and have been woven into ropes to allow for climbing activities.

The students and professor have designed a shelf-like device that lets an ape move across the top portion of his cage in an exciting variety of ways including log hopping, rope handling, ladder climbing, and overhand swinging.

The students have also experimented with aluminum and fiberglass devices similar to human pole-vaulting equipment.

CHAPTER V

Feed Manufacturer

Jerry Wilson's high school counselor tried to convince him that someone raised in the city couldn't possibly end up with a career that had anything to do with animals. Somehow, that just made Jerry more determined. He had loved animals since he was five years old and was sure there was a future for him in the animal industry.

He took lots of science courses plus the vocational agriculture course in high school. The agriculture teacher tried to discourage Jerry, too.

Undaunted, Jerry went to college. He received a degree in animal science and went on to graduate school to receive a master's degree in animal nutrition and reproductive physiology.

"I thought if I could learn how to feed and breed animals, I'd know the most important aspects of raising animals," says Jerry.

Jerry's first job after college was as a 4H farm adviser working for the California Extension Service. His task was to help organize the efforts of 300 4H leaders who worked with 1,300 youngsters. This experience developed in Jerry a great respect for the 4H program and its ability to develop leadership in young people.

Jerry reports that the Extension Service offers good job opportunities for young college graduates and that those positions often open new avenues for other jobs in the industry. Those new avenues led Jerry to a job as nutritionist with a middle-sized feed mill.

A nutritionist formulates animal diets and is involved in the quality control of the products. Because of the needs of his company, Jerry also became a salesman.

"I had been formulating the products and knew I had something good to sell," recalls Jerry. "I knew the product line as well as anyone, and I was permitted to set the prices almost as low as I wanted. Still I couldn't seem to give the stuff away. Selling was a whole different kind of thing. So I went to school to take a course in salesmanship. I learned how to close a deal and what motivates people to buy."

Jerry's successes in both sales and nutrition led him to his current position as plant manager of O. H. Kruse Grain and Milling, Inc., in El Monte, California. The mill, which employs sixty people, produces feed for layers (chickens that lay eggs) and specialty feeds for everything from mice to elephants, including horses, rabbits, and dairy cows. The company also exports feeds to foreign countries, such as hog concentrates to Hong Kong and Guam.

The mill is fairly old, with tall wooden cooling towers. In spite of frequent cleaning, the floors, pipes, conveyor belts, and ceiling are often covered with grain dust from the grinding and mixing processes. Hundred-pound sacks of grain mixes are stacked at the loading dock waiting for shipment by train or truck to dealers throughout the southwestern United States. Other feed is shipped in bulk. It pours through chutes into waiting open-topped truck trailers. To keep rodents under control, a problem for any feed producer, traps are set and cats are welcome unpaid employees.

The sixty people who work at this mill perform a variety of jobs, many of which might be found in any industry. There are secretaries, order takers, truck drivers, truck mechanics, production workers, machinery operators, feed mixers, and salesmen.

"Computers and computer programmers and operators are very important to our business," explains Jerry. "About 90 percent of the feed formulas are determined by computers. Particularly with layers, we pretty well know the requirements for vitamins, minerals, and aminos, plus the need for what we call 'unidentified growth products' that make chickens grow faster and bigger, although we still don't know quite why. We can plug all of that information into the computer, make some adjustments, and know what mix of feed will be the most economical for the farmer in terms of price per pound gained."

If Jerry is presented with a particularly difficult or unusual feed problem, he has access to nutrition consultants, many of them college professors or researchers, who have specialized knowledge of the feed business.

Since Jerry is plant manager, he must be concerned about many things beyond just producing good feed. His biggest worries revolve around transportation, weather, and government regulation.

"We get most of our corn and soybean meal from the Midwest or Texas," explains Jerry. "All of that has to be brought in by either truck or rail car. The transportation industry keeps jockeying for position, changing rates and capacities. You have to stay on top of the changes or your costs will soar."

Operating near the Los Angeles harbor gives Jerry's company one advantage. Many fish waste products are now used in feed preparation, and transportation costs within the area are relatively low.

Weather is the fickle foe of farmers and feed manufacturers alike. A few years ago California experienced a drought. The lack of water, and its high cost, caused many farmers to plant cotton rather than alfalfa. This shot the cost of alfalfa sky-high, and alternative feeds had to be used.

Although Jerry has no argument with the federal government's regulating his industry, he is concerned that there is a great deal of duplication of effort among the bureaucratic organizations that control his manufacturing facility. Much of his time and effort are spent trying to meet the requirements of the Federal Drug Administration while avoiding trouble with local smog-control authorities.

With all of the difficulties and daily problems to solve, Jerry still likes his job.

"I know lots of people would think the feed business is boring," says Jerry with a delighted twinkle in his eye, "but for me it's terribly important and exciting. It is always changing and challenging."

Jerry is very glad he ignored the advice of his high school counselor.

SOME FACTS

Feed is the largest single item of expense in the production of meat, milk, and eggs, accounting for 50 to 75 percent of the total costs. The feed industry ranks among the top twenty manufacturing industries in the United States and is the largest industry serving the American farmer.

In the last fifty years the feed industry has changed considerably. In the 1930's a major feed manufacturer offered 25 to 40 basic feeds. Now that same manufacturer would list up to 130 different basic formulations for all classes of livestock and poultry, plus specialty feeds for pets, rabbits, fish, birds, mice, monkeys, and other animals. Each product may be offered in several forms, including meal, crumbles, pellets, blocks, and liquid. It is possible for one manufacturer to have approximately 800 different finished products, including those with special additives, available to dealers and their customers.

The feed industry employs nearly 200,000 people. The number of employees per firm ranges from a few to about 20,000 employed by the largest of the manufacturers.

In large companies a person's job tends to be specialized, whereas in a small company a person may be asked to perform several tasks. For example, the nutritionist of a smaller concern would also handle feed registration, formulation, and quality control.

The total demand for food and feed is increasing substantially throughout the world with the continuing population explosion. The future of the feed industry looks good.

CHAPTER VI

Animal Science Teacher

Clarence Mann has been interested in animals and landscaping since he was a teenager. During high school, he took all of the available agriculture courses. His biggest project was raising pigs, which he exercised to keep them fit on the school's running track, keeping pace with the student athletes.

Though his first love is landscaping, he majored in animal science at college. "It didn't seem at the time that there was much future in landscaping," recalls Clarence. "There are certainly more opportunities now in landscaping, but it is still very competitive."

Clarence decided to become a teacher, which required a fifth year of college, for a variety of reasons. "I was never a particularly good student in high school," he reports, "and some of the teachers gave me a hard time, never gave me a second chance. I resolved that if I became a teacher, I wanted to work with those kids who maybe aren't so bright or outstanding. I still relate best to that kind of student, and it gives me a really good feeling to know I've helped youngsters who otherwise might have dropped out of school." Clarence was also attracted to the teaching profession because of the security and fringe benefits that go with it.

Twelve years ago Clarence was given the task of starting an agriculture program at Narbonne High School in Los Angeles. The acreage he was given to work with, at the far end of the campus, was a virtual desert. Fortunately, local feed store owners and nurserymen were very helpful. Now there are landscaped areas, a greenhouse, chicken and rabbit hutches, and about twenty-five large meat animals. Some of the animals belong to the school and are used as breeders; others are student projects.

Teacher Clarence Mann visits with his flock.

"Our students are encouraged to have a money-making project, not just raise pets," explains Clarence. "The students borrow money from a bank to buy the animal. They keep track of all the financial records, the costs of raising the animal, and then sell it at auction. The projects can include raising chickens for fryers or layers, rabbits for meat and pelts, sheep, or beef cattle. Students may also raise rats for research animals."

With so much land and so many animals to manage, Clarence is really a small businessman. He has to worry about ordering feed, keeping fences in repair, paying bills, scheduling trimming and fertilizing of landscaped areas, and being sure the manure is properly disposed of. "I certainly have a lot more responsibilities than, say, an English teacher," he concludes.

As a teacher, he offers courses in exploratory animal science, forestry, landscaping, floriculture, and a vocational landscaping program. Students in the exploratory animal science course learn the basics from textbooks and also have an opportunity to learn to handle the various animals, including feeding, cleaning, dressing out (butchering) chickens and rabbits, and, during lambing season, castration and docking.

Since his students have projects that are typically shown at fairs and he is involved, as a leader, in Future Farmers of America, Clarence spends a lot of weekends away from home. About ten weekends a year, Clarence finds himself on a school bus going to fairs and contests with his students.

He also has responsibilities during school holidays and summer vacations. After all, sheep need to eat all year round, and plants need watering. Fortunately, he is allowed to hire one adult aide who does much of the general maintenance around the mini-farm, and he employs a student aide for weekends to check out feed for student projects and generally monitor the area.

"More than anything else, I think I am teaching work habits and responsibility," says Clarence. "Students may change their career goals, but here, partly because I'm a strict disciplinarian, they learn how to work and that makes them a better person."

If Clarence were teaching agriculture in a more rural area, probably many of his students would work on their animal projects at home, since they would have more backyard room, and he would concentrate more on theory in class. Then he would have to use some of his after-school hours to visit his students' projects.

Some of the problems Clarence has as a teacher are the same as those experienced by any teacher. There are pressures from unhappy parents and from school administrators who don't seem to understand the need for vocational training. There is never enough money to do the things he would like to do. There are increasingly frequent problems with vandalism that can have a disastrous effect on animals or plants. Since all of his courses are electives, he has to design courses that will attract students or he would be out of business.

Some of the problems are unique to the animal business. For instance, not too long ago twin lambs were kidnapped from their pens. Fortunately, the community came to the support of the school, showing considerable concern. The small local newspaper printed the story on its front page. A woman who had read the story heard the bleating of sheep in a neighbor's yard and called the police. It was a happy ending when the sheep were returned to the high school.

There have been other rewards to the job, too. His program has become a showplace that is visited by representatives of school districts from around the world and is a popular field-trip destination for younger children. His students have had an opportunity to landscape for community public buildings and have been recognized for their quality work. And Clarence himself will have an opportunity this year to represent his large school district in Washington, D.C., at a conference on vocational education.

The future looks secure for young people who choose to become agriculture teachers. A California college reports that all of their graduates who receive a teaching credential in an agriculture specialty can be placed and that there continue to be unfilled openings. Clarence suggests several reasons for this need for agriculture teachers while other teaching positions are hard to find. "There is a high turnover among agriculture teachers," he suggests. "You have to work long hours and be away from home more than other teachers. That requires a very understanding wife. Some teachers also find that they can make more money by owning a farm themselves. Sometimes they teach and start small with a farm. When it becomes self-supporting, they quit their teaching job."

Most high school teachers' starting salary is about $10,000 a year. The current top salary is about $21,000. There is generally extra pay for summer responsibilities. Like many jobs that used to be strictly for men, more and more women are entering the profession. In the Los Angeles School District, for instance, twelve of the fifty agriculture teachers are women.

IF YOU LIKE DOGS

CHAPTER VII

Show Dog Handler

"Showing dogs is addictive," says Joe Waterman, who is one of the top ten handlers in the United States. "Once you get involved, it is hard to give it up."

Joe first got involved in dog shows when he purchased a Bedlington terrier, a middle-sized dog with soft white fluffy hair. It was sold to him on the condition that he show it. And so a hobby began.

"We showed her first in 'puppy matches,'" recalls Joe. "She wasn't a particularly good dog for show purposes, and that's not unusual for people with their first dog. So we bred her and kept one of the pups for show. I got more proficient at grooming, and other people began asking me to groom their dogs and show them. That became a training ground for me as I learned about the different breeds, how to groom them, and about their special characteristics.

"One year we campaigned a dog (showing a dog frequently so that it may earn enough points to be called a champion). That got very expensive, and I decided to start charging for my services."

One quick look at the show dogs in Joe's kennel and it is easy to see why he wanted to spend his full time with them. The dogs barely resemble the dogs seen in a normal neighborhood. Each dog has the distinctive lines of its breed and is groomed to absolute perfection. The first time you see show dogs, you realize that until then you have only been looking at imitation dogs. These are the real things—the Rolls Royces of the canine world.

So since 1967 Joe has been a full-time professional handler. It is not a hobby anymore. It is hard work.

There are 95 shows across the country each year, most of them on weekends. Sometimes as many as three shows occur on Friday, Saturday, and Sunday in neighboring towns. Joe shows dogs so frequently that he gets only one or two weekends off a year.

Joe doesn't own the dogs he shows. He is hired by the owners. They pay him $40 each time he shows the dog, plus transportation fees. The dogs stay at Joe's kennel while they are campaigning.

"Most of the dogs I handle are coated dogs rather than the smooth-haired type," says Joe. "You can't let a dog play with the kids and

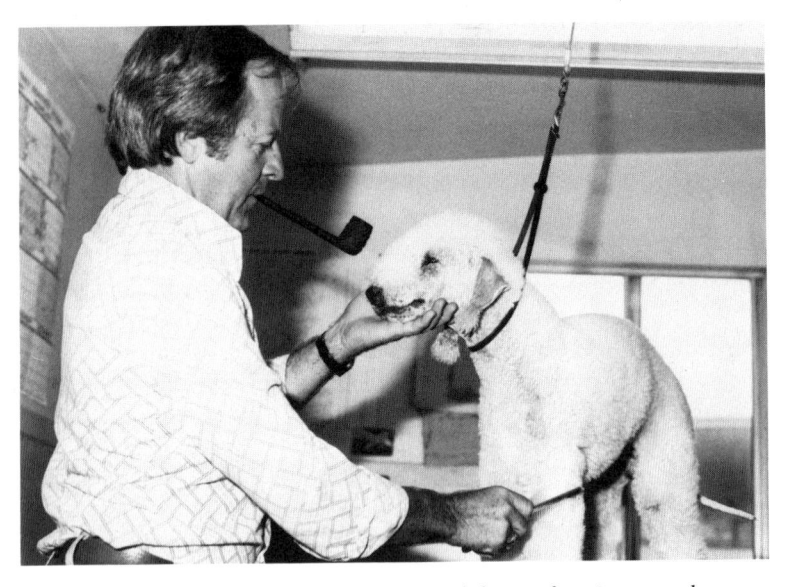

Bedlington terriers may have to be groomed three or four times a week.

run around outside and expect it to be in top shape. Some of these dogs have to be groomed three or four times a week. So we keep them here at the kennel all the time."

The night before a show, the activity is frantic for Joe and his staff of six, including his wife and one live-in employee. Every dog that is to be shown must be carefully groomed. Then they are loaded in their travel crates (wire cages) and put aboard the van. Joe, his wife, and one employee make the trip with the dogs. Usually he shows ten to twelve dogs at a time, although at local shows he has shown as many as twenty-two dogs.

"So that I'm not competing against myself, I usually show only one dog, one bitch, and one champion in any one breed," says Joe. "I could show more dogs if I specialized in the smooth-haired types that take less last-minute grooming."

Usually arriving late at night in the show town, Joe checks into a motel while his employee sleeps with the dogs in the camper. They are all up by 5:00 A.M. the next day and on the show grounds by 6:00.

Each dog is given a little exercise and groomed once again.

"We wash the legs and whiskers, and there is usually some last-minute scissoring to be done," says Joe.

The almost constant grooming goes on all day as Joe's dogs are scheduled to appear in the ring.

After so many years in the business, Joe knows most of the judges and can often guess what particular features in a dog a judge will look for. But there are still surprises.

"We were taking a dog to an Arizona show," recalls Joe. "The judges in both the class and the breed categories were not good for this dog, but we were campaigning her and this show was important. Well, both judges got sick. The substitutes loved the dog, and she won best in show."

Not all owners and handlers are entirely honest in their presentation of their dogs.

There was a German short-haired pointer who picked up all the prizes because he would always point when the judges were watching. Finally someone discovered that his handler was stuffing a dead pigeon in his shirt.

In another case, there was a German shepherd whose ears always dropped in the ring. The handler had his wife walk by the ring with a bitch in season, successfully attracting the dog's attention.

Sometimes owners rattle keys or call softly to their dog while he is in the ring, causing the dog to look very alert.

Every time one of Joe's dogs wins a category, it means more busy grooming for Joe to get the dog ready for the next category.

Usually the shows are over by 5:00 or 6:00 o'clock, and Joe can clean up and return to the motel for some hard-earned rest.

Fortunately, throughout such a busy day, there can be a few moments for fun.

"One handler beat me in the Bedlington terrier class," laughs Joe about the trick he played on a friend. "When he wasn't looking, I switched my dog for his in his trailer. The guy had my dog up onto the grooming table before he realized it was the wrong dog."

And almost no show is entirely routine.

"At another Arizona show, the owners of the property forgot to turn off the automatic flood irrigation system," recalls Joe. "We all woke up the next morning to find our vehicles standing in three inches of water."

Joe also tells the story of an older woman who was walking her standard male poodle on a rather long lead. The male happened by a female poodle in season that was being groomed. In an instant, the male had mounted the table and then the female. A great battle followed, with the owner pounding the young groomer with her purse while the groomer was trying to stop the male from doing what seemed to come so naturally.

The whole purpose of showing dogs is to make champions of them. The dogs earn points for the champion rating based on a very complicated system that depends, in part, on how many dogs they are competing against. Wanting a dog to be a champion is partly an ego trip for the owners and partly to make the dog more valuable for breeding purposes. In any one show, there may be 1,500 to 3,000 dogs entered. The reward for the owner is usually only a nice trophy.

After a busy weekend of shows, Joe returns to his kennel and the routine work of managing show dogs. His kennel can house seventy dogs. Eighteen of the dogs are actively in shows. The rest are former show dogs or ordinary dogs that have been kenneled while the owners are on vacation.

The show dogs are kept in individual wire cages in an air-conditioned building. They don't roam free at all. Everything is done to prevent the dogs from damaging their coats. Some dogs even have to wear panty hose to prevent them from scratching and causing the fur to fall out.

"You have to watch everything about the dogs very carefully," says Joe. "One flea can mean you won't be able to show a dog for six weeks, and that means no income from that dog."

Maintaining the dog's proper weight is always a worry. A two- or three-pound weight loss on a small dog can make him look all skin and bones. And that can happen almost overnight to a nervous male who is caged too close to a bitch in season. Or a Great Dane can lose three or four pounds if he gets homesick for his owners. A change of diet and extra tender-loving-care usually can correct this problem.

At the kennel, Joe spends much of his time on the phone with owners, handling problems and giving encouragement, while his employees take care of the routine work. Joe must also find time, however, for general maintenance around the kennel—fixing cages, repairing fences, and so on.

Grooming, however, is the key to success.

The show dogs, for the most part, are amazingly calm during the grooming process. It is on the grooming table that they learn to behave themselves for the show ring. Training starts early. Dogs can enter puppy matches at six to nine months. Adult dogs are shown starting around one year old, although larger dogs may not be fully muscled until age one and a half.

"Young dogs are just like juvenile delinquents," says Joe. "You have to give them the right discipline at the right time. Occasionally, I have a dog that is unmanageable. Then I just have to tell the owner he is wasting his money."

There are many local laws that control Joe's kennel operation. In his suburban community, zoning laws require that kennels operate only in manufacturing areas. Since Joe is on call 24 hours a day if there is trouble, he must live at the kennel. Although his home is quite attractive, his neighbors are other kennels or businesses.

Even so, the law says that dogs must be in an enclosed building after 7:00 P.M. to prevent disturbing neighbors with barking.

Joe suggests that young people who would like to become professional handlers apprentice themselves to a good kennel for at least two years to learn the business. They will probably have to work those two years at minimum wage.

During that time a young handler can begin to make contacts with potential customers. Once he breaks off on his own, he can expect very low income for the first year or so until he is established. It is probably wise to have some other source of income to tide him over this difficult period. In order to pick up more business, he will have to be regularly showing his dogs or those of other people.

A successful dog handler has a potential income of $20,000 or more per year.

CHAPTER VIII

Groomers

Snoopy is a reluctant visitor to the grooming shop. Not at all like the cartoon character, this Snoopy is a middle-sized, black, mixed-breed three-year-old. His ancestors might have included poodles and spaniels and who knows what else. At any rate, this is his first visit to any grooming shop. He tugs at his leash, trying to remove himself to more familiar environs. His tail is tucked between his legs, and he begins shaking from fear and nervousness.

Janice Worley, the groomer assigned to Snoopy, reaches down to greet him. He snaps at her, not out of anger but because he is afraid.

"It's all right, Snoopy," Janice says as she kneels so that she is nearly at his eye level. Her voice is calm, reassuring. "We're going to make you look so pretty. Oh, you're going to feel so good."

Janice takes the leash from Snoopy's owner and leads the still shaking dog into the back room. There, two other groomers are working on dogs who occasionally yap or whine. Janice lifts Snoopy onto her grooming table, a metal table about three feet wide by four feet long. She takes off his leash and attaches another leash that hangs from the ceiling. She is reluctant to put a muzzle on him, even though he has snapped at her. She continues to talk to him. It doesn't matter too much what she says. He hears only the tone. Slowly he calms down and stops shaking, although his tail is still between his legs.

Having quieted the dog, Janice is ready to begin the grooming process.

Three years of no grooming have taken their toll on Snoopy's wavy coat. The hair is badly matted. There is no way Janice can comb through those tangles. She begins with an electric razor, stripping away the fur on his back, leaving about a half inch. Great gobs of matted fur hit the floor in lumps.

After the worst of the coat is removed, Janice starts work on the long hair around Snoopy's face, which won't be cut so short. She brushes with a metal brush, separating the hair into small sections that can be handled without hurting the dog. The tangles are tough. Then she tackles the long hair of the tail. More matting is painstakingly removed.

Janice has to clip the toenails, too, the most difficult part of the job, and clean the ears.

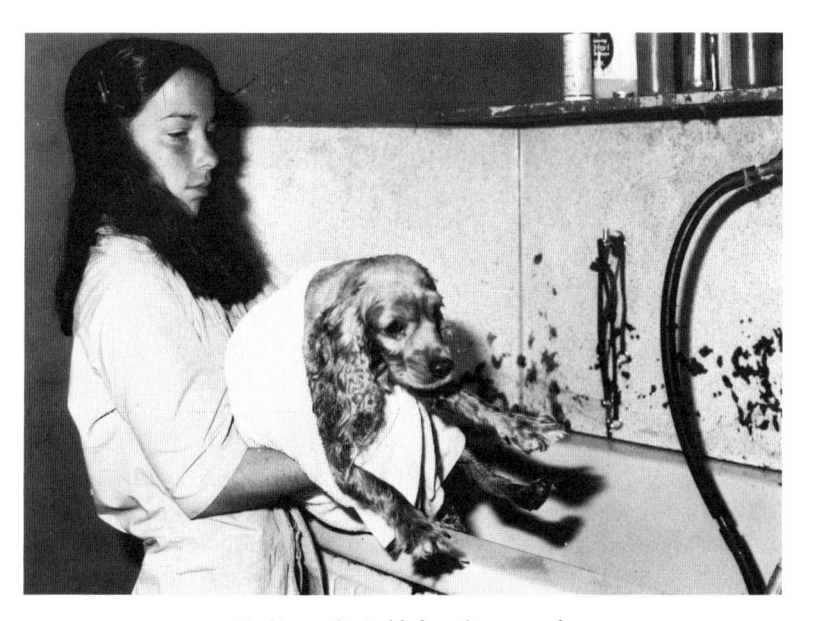

Bathing a dog is likely to be wet work.

Once this rough, or preliminary, grooming is completed, it's bath time for Snoopy.

Fortunately, most dogs like a good dip in the tub. It's wet work for the groomer, but at least the dog enjoys it. Then, with a series of shakes that spray the room with water, Snoopy is placed in a drying cage where warm blowing air speeds the drying process.

After Snoopy is dry, Janice turns her attention to the finish grooming process. Rather than a razor, she now uses scissors to trim and shape the final product.

Snoopy, when his owner returns to pick him up, is almost unrecognizable. He senses his sudden beauty. He prances and shows off his new hairdo. He even smells good. It takes some time to convince the owner that this is really his own Snoopy, even though the dog is jumping and leaping to show his long-time friendship.

Finally the owner, still shaking his head in disbelief, claims his dog, and a delighted Snoopy races out the door to the waiting car.

Janice, smiling at the beautiful dog she created from a mutt, returns to the grooming room and her next challenge.

Grooming, as an industry, owes its life to poodles and their increasing popularity over the last twenty years. Vern Burningham and his wife Beverly (now deceased) have been a part of that growth. Their successes are recorded on the walls of their office with blue ribbons and photos of their champion dogs filling most of the available space.

"I've been associated with dogs just about all my life," says Vern. "I started in Utah thirty years ago doing obedience training and training hunting dogs. After we were married, my wife and I worked very closely together. She was the national pet grooming chairman for the National Retail Pet Supply Association and conducted many workshops and seminars for that group. She was the first dog grooming instructor at any correctional institution. She taught prison inmates at the men's facility at Chino, California, and the women's facility in Frontera, California. She had such a vast knowledge of grooming and dogs that people frequently came to her for advice. She wanted to pass on her knowledge, so she founded this school, the Whittier Dog Grooming Academy."

Vern's school, one of the few schools scattered across the country, teaches grooming in a setting that is very much like a grooming shop. Up to twelve students are enrolled at one time. They are supervised by three experienced grooming teachers.

Students start off on the easier tasks: combing and brushing the dogs. As they gain experience, they learn to clean and pluck ears, cut and file toenails, and bathe dogs, including medicated baths. In the final stage of training, students learn scissor and blade finishing.

The course is designed so that students have an opportunity to work on a variety of dogs and at least one cat.

"We serve a clientele of 4,000 people," says Vern. "We make it a point to see that each student can handle any job when he or she graduates."

The course is designed for students to attend class eight hours a day, five days a week for fifteen weeks, or a total of six hundred class hours, most of it hands-on practice rather than lecture.

In addition to the cost of tuition, students purchase their own tools—scissors, blades, razors—an investment of about $350. These tools the students keep, to be used when they find employment.

"Grooming is physically hard work," explains Vern. "Combing a matted dog takes strength. You do a lot of lifting of dogs, and some of them are quite heavy. Probably the most difficult part of grooming to learn is the trimming. That's where the skill and artistry come into the job. It's almost like sculpturing the dog."

By the time the student graduates, he or she is capable of grooming six dogs a day without any errors. Then it is time to find a job.

Most grooming shops are small operations, employing no more than four groomers, and in some cases the owner is the only employee. The groomer works, as a beautician does, on a commission basis, usually receiving 50 percent of the charge to the customer. Starting off, a groomer can expect to earn only $25 to $35 per day. After gaining experience, a groomer should be able to do nine or ten dogs daily, and the income will increase if the shop has enough customers to maintain that level of performance.

In most shops, all of the day's dogs arrive in the morning. Other shops work on an appointment basis, with dogs arriving on schedule throughout the day. The dogs wait in cages for their turn at the rough trim. While one dog is in the drying cage after his bath, the groomer begins work on the next dog. When the dogs are finished, including a little cologne and a ribbon in the hair, the owners return to claim their pets.

Clearly, owners of grooming shops or pet stores with grooming facilities make more money than people who are just employed as groomers. Janice, like most young groomers, would like someday to own her own shop.

"It takes about $3,000 to open a grooming shop," says Vern. "The most important thing is to select a store on a well-traveled street but not in a place like a shopping mall. There has to be parking for your customers right at the door so they can easily drop off their dogs and even walk them around a little before and after grooming. The $3,000 should cover the first and last month's rent plus the utility charges and the cost of cages, dryers, and a showcase for the reception area.

It is a good idea to invest another $1,000 to $1,500 in merchandise to display in the reception area—impulse items like flea spray, leashes, and collars. The profits from those items should pay the rent."

Those who go into business for themselves should also realize that it may take a full year before they can build up sufficient clientele to make the business self-supporting.

Though being a groomer may be hard work, Janice is certainly happy with her career. She likes transforming a dog into a work of art.

"I've always loved animals and wanted to work with them in some way, to help them," she says. "I like this job because you can see the results right away. There is certainly more to the job than most people realize. What is really important is that pet owners become educated to the fact that their dogs should be groomed regularly for health and well-being."

CHAPTER IX

Police Dog Handler

Eight hundred people mill around outside a public building where a foreign political figure is meeting with local leaders. The crowd is made up of demonstrators who oppose the foreign government. They want to have a confrontation. They want to get publicity for their cause.

The television news cameras arrive on the scene. Now is the time for action, say the leaders of the crowd.

The crowd surges forward, waving posters and banners. Some protesters are armed with sticks, rocks, and bottles. There are probably unseen guns. The crowd knows that if there is bloodshed, they will receive more news coverage; more people will hear their story.

Waiting on the steps of the building are Dascha, a five-year-old German shepherd, and her handler, Officer Chuck Ferguson. Working with this pair are three other dogs and seventeen police officers. Eighteen men and four dogs against eight hundred angry protesters. Poor odds?

Chuck quietly gives Dascha the command, "Watch 'em." Dascha races forward the full length of her 20-foot leash, barking, growling, and baring her teeth. She sweeps right and left in an arc, threatening the crowd. The crowd moves back. One man is nipped when he doesn't move fast enough.

Chuck and the other officer–dog handlers move forward with their canine partners. The demonstrators at the edge of the crowd lose all interest in their cause; they return to their cars and leave. The crowd gets smaller. The TV cameramen lose interest, too. There is no excitement here.

Chuck calls Dascha back to heel and puts her on a short leash. "Good girl," he praises her, rubbing her chest. "Good girl."

Dascha is part pet and part aggressive police dog.

That evening Dascha is at home with Chuck. She is the family pet. Chuck's one-year-old son climbs onto Dascha's back and turns around to pull her tail. Dascha carefully lies down and rolls over, dumping the giggling toddler onto the soft rug. Tired from a busy day at work, Dascha stands, shakes herself, and strolls into the master bedroom, where she crawls under the bed to avoid, for at least the moment, her playful young friend whom she would in an instant give up her life to protect.

"There is a general misconception about police dogs," says Lt. Ralph Cook, head of Special Projects of the Inglewood Police Department in California. He supervises the four men in the department whose partners are dogs.

"These dogs are not mean, vicious animals," says Lt. Cook. "Their handlers always have full control of their dogs. But on command, these pets turn into very aggressive animals."

Police dogs are trained to respond, react, to abnormal or aggressive behavior. Handlers are also convinced that the dogs react to the scent of fear in a suspect.

Crowd control, such as the political demonstration, is perhaps the most dramatic use of police dogs.

"If I have to face down a crowd, they know that I, as a policeman, have rules to follow about using my gun or too much force," explains Chuck. "On the other hand, they know Dascha is not frightened. She is not bluffing. She will do anything she has been taught to do, without any fear whatsoever. So it is the deterrent factor, people's real fear of getting hurt, that makes dogs so effective in crowd control."

It is not unusual for just one officer and his dog to respond to a call at a local high school about some crowd disturbance. Generally, the crowd breaks up in a hurry and no one is hurt.

Dogs are equally effective for searching buildings for suspects.

"When I'm on duty, I don't have any particular assignment," explains Chuck. "I listen to the radio calls. If there is a silent burglar alarm, I go in as a back-up unit."

If the officers at the scene of the alarm decide the building should be searched, Chuck first announces that over a loudspeaker. "We are about to search this building, and we have a trained attack dog with us. Anyone in the building should come out at once with your hands up." At this point, to prove that he is not bluffing about the presence of the dog, he commands Dascha to bark.

Dascha, on a leash, and Chuck then enter the building.

"One of a new officer's biggest mistakes is not reading his dog, not having faith in him," says Lt. Cook. "The dog will try to tell his handler that there is something in a closet or behind a filing cabinet, but the officer won't understand. It takes a dog and his handler six months to a year to really begin working together as a team."

The presence of a dog saves lives and injuries.

"We sent a dog into a sporting goods store," recalls Lt. Cook. "The store sells guns and ammunition. Just the threat of the dog caused three would-be robbers to throw down their guns and come out, even though one of the guys had a military-type repeating rifle all loaded and ready to go. There is no question in my mind that if we had just sent in officers, without the dog, someone would have been hurt."

Sometimes dogs are asked to search a large outdoor area.

"Our officers interrupted a robbery in progress," says Lt. Cook. "The suspect took a shot at our officers, then got in his car and took off. Officers pursued him at high speed for several miles until the suspect ran his car into a telephone pole. Evidently he was unhurt, because he leaped out of the car and ran off between two houses. The dog and handler went after him. In ten minutes, the dog had found the suspect hiding under a house. He had used the crawlway to get in. The dog actually went in and dragged him out."

What could have been a very dangerous situation with an armed robber turned into an almost routine arrest.

Some dogs have special talents or training.

One of Inglewood's dogs is particularly good at tracking. He and his handler have been called to neighboring counties for special duties involving possible homicides as well as the usual duties of finding lost children and catching ordinary criminals.

Police departments capitalize on a dog's keen sense of smell by training them to search out bombs or narcotics. Usually these dogs are not used in crowd control or building searches for suspects. "Cross-training seems to confuse most dogs," says Lt. Cook.

Often it is the airport security department, a very large police department or the border patrol that has a bomb-detection or narcotics dog. To train a dog for this sort of service costs $6,000 to $8,000 dollars. The expense for the other dogs is about $2,000 to $3,000.

"A really good narcotics dog will get the scent of marijuana hidden in, say, a couch and not just sniff at it but literally rip the cushions apart trying to get at the marijuana," says Lt. Cook.

Training a dog and its handler is no easy task, and it goes on constantly while the dog is in service.

"Dascha was about three years old when we got her," recalls Lt. Cook. "She had been a show dog and brood bitch. She had two litters. We had looked at several kennels that train guard dogs. The kennel we picked had furnished trained dogs to other police departments, seemed well run, and had an inventory of several dogs that were already trained."

The kennel furnishes dogs that are trained in agitation (the barking, growling, and baring of teeth), to attack on command, to protect their handler, and to respond to being called off. Most of the dogs are about two years old.

Selecting an officer to become a dog handler is a lengthy process, too. Chuck was selected from a list of twenty police officers who had volunteered for the assignment.

"We look for self-control and stability in the officers we pick," says Lt. Cook, "plus an ability to work alone in the field."

Once selected, Chuck was assigned to two weeks' duty at the kennel with his new dog, Dascha. It took ten hours a day in training, and Dascha and Chuck lived together.

"It's here that the bond is formed, the handler/dog relationship that is so important to a successful team," says Chuck.

During those two weeks, Chuck was trained in basic animal care and handling as well as the techniques that apply to police assignments.

Most of the officers who apply for this duty do so to broaden their general police experience and to provide some variety from the usual patrol duty. Among other things, they know that the dog handler is involved in a lot of arrests but usually has to do far less paperwork than do the patrol officers. And, like Chuck, very few of them have any great experience with animals except for family pets in their childhood.

On the job, Chuck and Dascha have to keep up their training in order to stay sharp.

"Every day I spend 15 to 45 minutes with Dascha on general obedience training," says Chuck. Four or more hours a week, they work on attack training.

"If we search a building and don't find anyone, we may turn it into a training exercise," says Chuck. "One of the other officers present puts on the old jacket I keep in the car (dogs are never allowed to attack an officer in uniform) and the sleeve (the padded arm covering that is worn for the attack point). Then the officer goes back into the building and hides. We go looking for him, and Dascha is allowed to attack. This sleeve training is like a game to her. She really likes to bite. And if we are always searching places without finding anyone, she forgets why we are doing it, doesn't feel rewarded."

It would be hard for a young person to set his goal in life to become a police officer/dog handler, since very few police departments employ dogs. However, there is a trend toward greater use of dogs, since it often represents the minimum force necessary to accomplish a police objective. And, according to Lt. Cook, police departments are finding that police dogs save money for the city by reducing both police injuries and the number of officers that must respond to the scene of a crime.

CHAPTER X

Guard Dog Trainer

The two dark brown Doberman pinschers, silhouetted by the moonlight, trot along inside the fenced lumberyard, their sharp pointed ears alert for any sounds of intruders. A stranger approaches the fence. The dogs race to challenge the man. They bark angrily at him and snarl, drawing their lips back over their teeth. They follow the man as he walks beside the fence. Finally, the man crosses the street and returns to his car. The dogs resume their jog through the maze of stacked lumber. "Anyone would have to be nuts to mess with dogs like those two girls," says Kem Roddy, owner of the pair and dog trainer. Surprisingly, Kem confesses that the two dogs would probably not actually bite someone who entered when they were on guard duty. The dogs have not been trained as attack dogs. They have been "fence-trained" through confidence and courage-building to scare away anyone who is not welcome in the guarded area. They are very effective in their job.

"I try to discourage people and businessmen from having dogs that are attack- or sentry-trained," explains Kem. "If you use those kinds of dogs, someone is going to get hurt and it will probably be a kid. I have never had a piece of wood stolen or a window broken where I've had my dogs on duty."

Kem, only twenty-two years old, has owned the K-9 Protection and Obedience School for almost a year. He offers training in basic obedience, personal protection, problem solving, and attack, and he supplies or trains guard dogs.

Although Kem doesn't necessarily recommend military service, he received his training in dog handling in the Air Force.

"I enlisted with a guaranteed job in law enforcement," recalls Kem. "After I completed some of the basic courses in police work, I volunteered for dog training. I had several letters of recommendation from my commanding officer and other officers, and that helped me get into one of the classes at Lackland Air Force Base. There were only fifteen students in each class."

During the three months of intensive training with dogs, Kem learned basic obedience, how to conduct field and building searches, and arrest procedures. He also learned riot control with the use of dogs. He is very proud of the training and knowledge he has gained.

"It was a very prestigious job," he says.

It was also a potentially dangerous job. Kem patrolled dark, unlighted areas with a single dog. He searched empty buildings looking for vandals or thieves. He learned to rely heavily on the keen senses of his dog to protect him from the unexpected.

Now, as a small business man, Kem's day starts about 5:00 A.M. as he picks up his guard dogs from various locations. Businesses pay him $25 per night for each dog that works.

After a quick breakfast, Kem begins working with the dogs that he has been hired to obedience-train. His fee is $200. The dogs are kenneled nearby (until Kem can afford to build his own kennel) and are returned to the owner in two or three weeks fully trained.

"When I meet a dog for the first time, I spend a lot of time just getting acquainted with him," explains Kem. "You have to understand the dog's personality and what will work well with that dog. Maybe in the first half-hour session, I'll spend fifteen minutes just playing with the dog. For the most part, you treat dogs like three- or four-year-old children. You talk to them like you would a kid. They don't understand the words, of course, but they get the tone of voice."

Usually Kem uses a pinch collar for training purposes. This is a chain collar with links that pull together to pinch the neck skin when the leash is pulled. There is a small amount of pain for the dog, enough to get his attention but not seriously hurt him. Sometimes, with a particularly sensitive or fearful dog, Kem uses a choke chain that simply becomes very tight when the leash is yanked. The real secret of Kem's success, however, is love and praise.

Sometimes problem dogs are brought to Kem, dogs that dig holes in flower beds, chew furniture, or tear up things.

"Think of the worst three-year-old child you know, a spoiled little kid, and you'll know what a problem dog is like," says Kem. The same two or three weeks of discipline and training usually solve the problem for the owner.

During the afternoon hours, Kem does the necessary paperwork that goes with any business. He has about $1,500 to $2,000 worth of bills that must be paid each month. His expenses include want ads in the telephone yellow pages (his single best source of customers), telephone costs, veterinarian bills, kenneling expenses, and $300 worth of dog food. He is very pleased that even in his first year of ownership he has been able to pay all of his bills and have enough left over for living expenses and a new motorcycle.

Late afternoon finds Kem dealing with more obedience-training customers. He is planning to start a course for prospective dog handlers. He will charge about $2,000 for a several-month course. Some other businesses charge as high as $4,500 to teach a person to be an obedience trainer.

"I have so much knowledge about handling dogs that I really feel I have to share it," says Kem.

As darkness approaches, Kem puts his guard dogs back in the truck to deliver them to their assignments. That task marks the end of a long day that has included, in spare moments, the tasks of feeding the dogs, cleaning kennels, and grooming some of the dogs.

Kem prefers to train dogs that are nine months to a year old. If he needs an additional guard dog, he either buys a dog of that age or, through an ad in the newspaper, "Save a Pet Homefinders," he locates Dobermans or German shepherds whose owners cannot keep them anymore.

When Kem approaches a potential customer for his guard dog service, he has some important sales points to make. A dog is always alert on duty, never sleeping or reading; a dog has a better sense of smell and hearing than any man could have; a dog senses danger faster than a man; a dog works as many hours as necessary; a dog has no personal problems that will interfere with his work; a dog gets no fringe benefits like vacations and days off; a dog works every day of the year; and a dog has a psychological advantage over an intruder.

There are some surprise hazards in this business. Kem has been threatened, presumably by a competitor, that if he didn't go out of business he might be hurt. The prior owner of the business reportedly had at least one shot taken at him through the bathroom window of his home. Kem doesn't seem particularly worried.

"I can't think of any career field that would be more exciting," he says. "The money is unlimited for those with creativity and energy."

Kem has made a good beginning on an original investment of $2,000, lent to him by an uncle.

CHAPTER XI

Guide Dog Trainer

Intelligent disobedience. That is what Al Whitehead of International Guiding Eyes looks for in the dogs he trains to lead blind masters safely through their daily activities.

Most guide dogs are German shepherds, although golden retrievers and Labrador retrievers are also used. Female dogs are used because their temperaments are usually a little better suited to the work and their personal habits are better than males.

Al gets his dogs from several sources. Some are born right in the kennel. Others come from breeders who may find a particular dog is not suitable for show work but would be quite all right as a guide dog. Still others are donated by families who for some reason can no longer keep the dog.

Al keeps the puppies at his kennel until they are three months old. Then they are placed in a family home. 4H Clubs across the country are very much involved with raising guide dogs.

"The dog lives in the home as a family pet for about a year," explains Al. "She is taken to the beach, into stores, along streets, and generally socialized. Through the structured program with 4H, we are able to get the dogs exposed to construction sites and the loud noises there; take them to farms where they can develop a respect for, but not fear of or a desire to chase, the farm animals; and experience some rule-setting. There are usually group obedience classes for the dogs. We can even arrange for the dogs to ride buses to get used to the noise and motion. Where people are raising the dogs not in a group, we don't have as good control over the experiences of the animal."

When the dog is just over a year old, she returns to Guiding Eyes for five months of intensive training to become a guide dog. About 30 percent of the dogs fail the training.

"Some dogs are too shy of people," explains Al. "Others are frightened of noises or other things in the environment. The dogs need to have some initiative. They have to be able to decide when they come to an obstacle whether to go around it to the left or right, without their master giving any commands. Some dogs just aren't sensitive enough, or they may be hyperactive. A guide dog must be a very responsible animal."

First the dogs are taught general obedience—to come, sit, fetch. Then they are introduced to the harness, how to lead and pull. It is through the harness that the blind master feels the direction the dog wants him to go or when he should stop.

Then the dog learns about curbs. He must learn to stop at each curb, even if it is one of those sloping curbs intended for wheelchairs, and wait for instructions to continue.

It is also important for the dog to learn to cross streets in a straight line. If the dog crosses diagonally or wanders out into the cross-traffic lanes, it would be dangerous and confusing for the master. And the dog must learn to intelligently disobey his master if told to go forward when a car or truck would be a hazard.

The training is begun in quiet residential streets. At first the trainer shows the dog what is expected of her. Later the trainer is blindfolded and the dog actually leads him, with another trainer walking behind to warn of any dangers the dog does not pick up.

After about three months of training, the dog is introduced to heavier traffic situations, then to other pedestrians and obstacles along the route. She has to learn to go around objects and avoid people. She also must be aware of people, even those on bicycles or skateboards, who may be overtaking her. Hopefully, those who do pass a blind person with a lead dog will be courteous enough to allow the dog plenty of room and will not interfere with the progress of the blind person.

The hardest thing to teach a dog is to be attentive to what is above her, the overhanging obstacles that might hurt her master but are no problem to her since she is only two feet tall.

"Fortunately, we don't encounter many overhead obstacles," Al points out. "If there are obstacles on the regular routes the blind person travels, the dog learns to avoid them. Elsewhere, she is likely to miss it. It is a weakness with these dogs."

Training a guide dog costs $4,500 to $5,000. The dogs are given to the blind master at no cost. Charitable contributions absorb the entire cost.

Blind people from all over the world apply for a guide dog to one of the eight training schools in the United States. Some of the schools have waiting lists as long as a year.

"We personally interview all of the applicants for guide dogs," explains Al. "Not all people can work well with dogs. For instance, a person is legally blind if he has 20/200 vision. If the person has some sight, he may give the dog confused signals and it will not work out well. The person has to be physically and financially able to take care of a dog. If the person rents his house or apartment, he must have permission from the landlord to have a dog. His doctor must also approve the person as being suitable for a guide dog. A person who is hard of hearing would not be suitable, since he would not be able to hear the traffic sounds. Most important, the person must want the dog to improve his independence. It's a terrific responsibility to take care of a dog."

The blind student must spend a month in residence at the Guiding Eyes facility learning to operate with his dog under all circumstances. Up to sixteen students can be trained at once, each of them matched carefully for personalities with their new dogs.

The blind master goes through very much the same kind of training that the dog has already completed, except that the master is in charge. The student learns to listen for traffic when crossing streets by hearing the traffic patterns. He may have to count blocks or know the names of streets as he crosses them, feeling for landmarks to assure himself that he is moving in the right direction. He has to listen for one-way traffic, cars pulling out of driveways or turning right on a red signal. When the dog stops at a curb, the master searches for the curb with his foot and then commands the dog with both a hand movement and voice to go forward. Most of all, the student must learn to trust the animal that is helping him to regain his independence and to walk with his body erect and his head held high.

Twice each day, student, dog, and trainer work together. The dog and her master do not go for quiet little strolls. They are taught to walk at a very rapid pace.

"It's actually easier for the blind person to keep his balance if he walks rapidly," says Al. "And it is better exercise. If you go too slowly, the dog loses interest and forgets she is working."

Some of the students work under an added handicap. They have

Apprentice guide dog trainer Diane Gomsett learns how it feels to be blind. Instructor Al Whitehead shows her how to hold the guide-dog harness.

never had a pet of any sort. It may take them added time to build the confidence necessary to make the dog and student a good team. By the end of the first week, most student/dog teams have become close companions and can communicate in a very special way.

To become a guide dog trainer takes three years of apprenticeship training. Because there are so few schools in this country, opportunities are limited. Pay rates are not high. Apprentices start at about $600 per month and may earn $1,000 per month when fully qualified.

"When I hire an apprentice, I like to see someone who has had at least one year of experience with dogs," says Al. "They might have gotten that experience in the military, working for a veterinarian, or helping in a kennel. At any rate, they should know something about the care of animals. If they haven't worked with animals, we accept course work in sociology or psychology in lieu of experience."

The apprentice learns how to care for the dogs, including grooming and kennel work. At first they learn to teach basic obedience and then progress to guide work. Part of that training includes putting on a thick black blindfold and getting a feeling for just what it is like to be blind and led.

Diane Gomsett, a new apprentice, recalls her first trip under the blindfold. "I was terrified. I thought I would stumble on a crack or something. And I was really afraid of the traffic."

For Diane's first trip as a blind person, Al became guide dog "Heidi." He held the harness at about the level Diane would feel if it were a dog. He showed her where to hold the harness, keeping it near her leg where the touch would be the most sensitive and the messages the most clear. At each curb, going up or down, Diane cautiously reached out with her foot before giving the command "forward," "left," or "right" plus the arm signal that goes with the command.

During Diane's apprenticeship, she will be required to read and understand several rather thick books on veterinary medicine, dogs, and the management of people. Both the reading and training will help her to pass state licensing examinations to become a certified guide dog trainer.

"The most difficult part of being a trainer," according to Al, "is learning self-control and self-discipline. Dogs and people learn at different speeds and levels. The job requires a great deal of patience. If you don't show patience, you will just frustrate the student and confuse the dog. The entire job requires diligence and attention to consistency."

Al views each of his students as unique, but perhaps the most unusual situation for him was the student who had an artificial leg. Because of problems of balance and feeling, the student would have to work his dog on the right side instead of the usual left side. Al had to train the dog from the beginning to work on the right, which meant that all of the hand signals were opposite and the positioning of the feet was reversed. Somehow, Al didn't seem to mind the challenge.

"We're here to provide a service," explains Al. "We help the individual gain more freedom and independence. You get a warm and refreshing feeling when your efforts pay off."

The American Humane Association in Denver, Colorado, has been involved in a project to train dogs and cats to hear for deaf people. The animals are trained to respond to the sound of door bells, telephones, and alarm clocks and to alert their owners. The project has always been small and the funding insecure.

Students interested in this career field may want to read *Banner, Forward, A Pictorial Biography of a Guide Dog,* by Ena Rappaport, E. P. Dutton and Co., Inc., New York, 1969.

LOVE OF HORSES

CHAPTER XII

Racetrack Trainer and Jockey

It takes nearly 3,000 people to care for the horses at a major racetrack. And the track employs another 1,000 people as ushers, guards, and bet takers.

Out of view of the daily race spectators, there are rows upon rows of stalls, enough to house 2,500 horses in the case of Hollywood Park racetrack in southern California. In one- and two-story buildings scattered around acres of grounds, horses share space with offices and living quarters of track employees. Many of the employees are Spanish-speaking, often from Mexico. During morning hours, clusters of jockeys exchange the latest track gossip while waiting for riding assignments. Exercisers, frequently jockeys of the future, train horses for speed and endurance. Groomers soothe the aching muscles of tired horses or clean stalls.

For the horses, day starts at 3:00 A.M. with breakfast. The night man, who works for a particular trainer and lives at the track, feeds each horse his first ration of the day. That meal must be reasonably well digested before the horse is asked to exercise vigorously during the early morning workouts.

By 5:30 A.M. the trainers begin to show up at the track.

Loren Rettele, just over thirty years old, has been a trainer for five years. Before that he was a groomer, exerciser, and assistant trainer.

"I was born and raised on a farm," says Loren, "so I knew a lot about horses and liked them. After I came out of the Army, I was going to work for the phone company, but I decided I could make a lot more money in racing."

And indeed Loren is making money as a trainer—over $100,000 in

1978—but the work isn't easy. It is a fourteen-hour-a-day job, seven days a week, and it is hard to get away for a vacation. But then, Loren really enjoys his work and beams with pride as he talks about his winning horses.

Loren doesn't own the horses he trains. Fourteen owners trust Loren's judgment and ability enough that they have placed thirty-four horses in his care. They pay $30 a day per horse for that service, which just covers Loren's costs of feed and the employees he must pay. (The horse stalls are provided free by the racetrack.) To have that many owners trust him means that Loren is not only a good trainer but also a good salesman. Loren really only makes money when the horses win and he collects 10 percent of the prize money.

Training a horse takes time and patience.

"Some horses I get when they are about one and a half years old," explains Loren. "Each horse is an individual and needs to be trained a little differently. You usually start horses on the shorter, six-furlong (three-quarter-mile) race. You train an older horse, four or five years old, harder than you do the two-year-old. And fillies you train easier. Some horses have a natural ability for the longer races, just as some people can run marathons whereas others are better at sprints."

Loren's best horse so far, Golden Act, has placed third in the Kentucky Derby and second in the Belmont Stakes, both long races, while earning $496,000 in his three years.

Each day, Loren plans that day's exercise for each horse. The daily chart tells the exercisers whether to give the horse a hard workout or just canter around the track to build endurance.

The exercisers start their work about 6:00 A.M. Loren employs four exercisers who earn about $600 to $1,000 per month, plus a percentage of the winnings. The exercisers' day is usually short—they are often finished with their work by 10:00 o'clock—but they can be pretty tired after four hours of hard riding. Most exercisers have been groomers for a trainer who gives them a chance to ride or have worked at small tracks around the country or on horse ranches before coming to a major track like Hollywood Park.

Sometimes a horse gets sore legs and should not be run. Those horses are taken for a swim. Snorting and paddling with all four legs, the horse exercises in a pool perhaps 30 feet in diameter, led around and around by a handler who walks beside the pool. Once used to the idea of swimming, most horses seem to enjoy this exercise.

Groomers, like the exercisers, start their work day at 6:00 A.M. They earn about $800 per month, and their hours are longer than those of the exercisers. They get the horses ready for the morning workout, clean and put fresh straw in the stalls, wrap legs as needed, and may give medication or vitamins. They also have to care for the saddles and bridles, seeing that they are kept in good condition.

When the horses are brought back from their workouts, they must be cooled down slowly. Some horses are harnessed to a mechanical hot-walker, a kind of merry-go-round that keeps the horse moving slowly in a circle until he stops sweating. Other horses are cooled by a human hot-walker who receives the lowest pay of the track employees, about $100 per week. The human hot-walker leads the horse around the stable area until he is cool enough to be returned to the groomer.

By 11:00 o'clock activity around the stable has slowed. Now the groomers feed their horses lunch, early enough so that digestion will not interfere with the horses' performance during an afternoon race. After lunch, the horses rest, many of them actually lying down to take a well-deserved nap.

There are nine races a day at this track. Loren's biggest job, and the one that pays off, is choosing which of his horses to run in which race.

"The object is to win, so you put a horse in a race against horses of similar ability," says Loren. "The Racing Secretary (who works for the track) decides what kind of races will be run."

Different kinds of races attract different caliber horses. The lesser or newer horses often run in "claiming races." In a claiming race for $8,000, any horse that runs may be "claimed," or purchased, for that amount of money by anyone who chooses to do so. A trainer will enter a horse that he feels is worth less than the $8,000, willing to risk the loss of the horse to get a winning purse of about $15,000. A claiming race of $32,000 draws somewhat better horses, and the horses are still faster in a claiming race for $60,000.

Loren may have only two or three horses running on any given race day. He carefully picks and chooses the races that will be best for his horses. Of course, all of the other trainers are doing the same thing.

Choosing the right jockey is another part of finding the winning combination. Jockeys, like movie stars, have agents. The agent's job is to get his jockeys the best mounts available, the horses most likely to win. Loren's job is to pick the best available jockey for his horses that day.

"There are six or eight really top riders," says Loren. "I'll take any one of those if they are available. The rest I just have to judge by their records, trying to get the best I can."

Most jockeys start as hot-walkers or exercisers. Or they may start at a smaller track and build a reputation as they gain experience.

"Some jockeys just stick with the small tracks and make a very comfortable living for themselves," Loren points out.

Sometimes an exerciser who has worked for a particular trainer for a while gets a contract with that trainer for one to three years as a jockey. That gives him a start and a record to build on.

There seems to be no limit to the amount of money a top jockey can make. The successful jockey, even though he may not be well known, is still likely to earn a handsome income of $60,000 to $70,000 per year if he can keep his weight down.

Willie Shoemaker, probably the most famous jockey, has always kept his weight around 100 pounds and done so fairly easily. Laffit Pinquay, on the other hand, struggles to keep his weight at 117 pounds including the two or three pounds worth of saddle and boots.

Jockeys, as a rule, diet constantly, take sauna baths to sweat off the weight, and exercise regularly to use up the calories.

"If a jockey starts off at age 17 or 18 weighing 105 pounds, he is likely to have trouble by the time he is fully grown," says Loren.

There is no question, however, that a jockey needs, in spite of his small size, strong legs, back, and arms. Racing is a demanding sport.

The horses are fed again around 5:00 o'clock, and Loren's work is usually done so that he can be home by 7:30, or somewhat earlier on Monday and Tuesday, which are not race days.

Loren learned his trainer's trade at Midwestern tracks. With the changing racing seasons, he would spend two or three months in each of several cities: Detroit, New Orleans, Chicago, Omaha, and Louisville. He married a girl whose father had been a trainer and who knew what to expect in the racing business. When their son was approaching school age, they decided to move to California where three racetracks, Hollywood Park, Santa Anita, and Del Mar, were all within driving distance. They could have a home and put down roots like nonracing people, and their son would not have his schooling interrupted. Now Loren only travels when he has a horse that can qualify for entry in the really big races like the Kentucky Derby.

Betsy Allen, unlike Loren, doesn't leave the racetrack at night. She is one of the thousand track employees who live there. Her room, which is furnished rent-free by the track, is about 12' x 14'. There is a single bed, a desk, a rod to hang clothes on, and a refrigerator, which she rents. A hot plate and a slow-cooking pot are her kitchen appliances.

Betsy has lived around the track for about three years. For a while she was a hot-walker and groomer. Now she works for the track's publicity department. She is thinking about going into the sales business—supplies for the racing or horse recreational trade.

"I feel that my future in racing is rather restricted," says Betsy. "I came to the racetrack because I love horses, but I can't see how I can make more than about $12,000 to $15,000 per year." Betsy knows that there are very few women trainers or even jockeys, although an increasing percentage of groomers and exercisers are women.

"Living at the track is nice in that it's like having a family around you all the time," says Betsy. "But you are limited in the kinds of people you have a chance to meet."

That same statement would be true of any seven-day-a-week job.

CHAPTER XIII

Blacksmith

Like horse trainers, exercisers, and groomers, Dave Richison, a blacksmith, spends his days at the racetrack. He is one of twenty-four blacksmiths at Hollywood Park in Los Angeles who specialize in the care of racehorses. He is the fifth generation of Richisons to earn their livelihood in this ancient trade. His grandfather learned the trade in Ireland.

"I spent five years as an apprentice working for other smiths," recalls Dave. "I learned basic horse safety and how to get along with the different temperaments of horses. For instance, you have to know when to reprimand a horse and how vigorously. When I first started, I would just catch the horses for the master smith. Then I learned to pull off the horseshoes and how to use the tools. Finally, I learned how to shape and nail."

Dave handled saddle horses for several years before he decided to work at the racetracks, a move that he hoped would help him earn more money. Before he could be licensed for racetrack work, he had to pass a six-hour written examination and a five-hour practical test. The testing turned out to be the easy part.

"During the first three months I worked at the track, I shod just one horse," recalls Dave. "On the way home, I would do saddle horses just so I could have enough income to eat."

It took Dave five or six years before the trainers had enough faith in his ability to provide really steady work. Since there is so much at stake with a racehorse, the long time required to prove yourself is not unusual.

In addition to the pressures involved because of the large investment each owner has in his horses, there are some special problems for the blacksmith. "When they breed racehorses, they often don't take the feet into account," says Dave. "There is so much inbreeding that these horses often have problem feet." Dave finds he has to create shoes that actually change the mechanics of how a horse runs so that he will not kick himself or run crooked. "A lot of the work is trial and error. You have to try what ought to work, and if it doesn't you try something else."

For the sake of reducing the weight a horse must carry, most race horses wear aluminum shoes.

Dave works six days a week. Sunday is his day off. The only holiday he celebrates is the Fourth of July. His vacation comes when the three tracks he services are all closed.

The day starts early for Dave: 7:00 A.M. He makes the rounds of the stables that have hired him as an independent contractor, checking with each trainer to learn if any of the horses have thrown a shoe or developed a problem. The horses have to be ready for their morning workouts. After handling any emergencies, he begins the more routine work of changing the shoes on horses. Most racehorses have their shoes changed every 21 to 30 days. Usually, Dave is finished with his work and back home by 4:00 P.M.

Of course, there is a certain amount of danger when working with any large animal. "These horses can kick, bite, or flip over backwards and fall on you," says Dave. "It's a business where it is hard to get medical insurance."

Dave suggests that young people interested in becoming blacksmiths attend a vocational school first and then try to work with another smith. "When you first get out of school, you get every bad horse and every nonpaying customer that the other blacksmiths don't want," warns Dave. "And even to get started takes $8,000 to $10,000 worth of equipment."

Not far from Hollywood Park, Tim Dierks works as a blacksmith, too. He works in an area called Palos Verdes Peninsula, a wealthy residential area with homes that cost $150,000 to $500,000 and more. Many of the homes have their own stables. Tim estimates that there are 10,000 horses living on the Peninsula and about ten to fifteen blacksmiths who work the area. Every blacksmith is very busy.

Tim is his own boss and likes it that way. He learned his trade from his father. He works six or seven days a week from 8:00 A.M.

to 6:00 P.M. He has a route of customers whom he serves on a regular basis. Most of the saddle horses are shod every two months, though some horses that are seldom ridden may not need shoes at all. Tim arranges his schedule so that he works very hard for the first three weeks or so of every month, and then he takes several days off at a time for a fishing trip to the High Sierra mountains.

The horses on the Peninsula are owned primarily for recreational purposes, but there are many kinds of recreation and each sport or activity requires a different kind of shoe. The English hunter-jumper horse needs a lightweight aluminum shoe. A polo horse wears a plastic or hard rubber shoe. The Western riding horse wears the familiar steel shoe. In each case, there may be special problems that Tim has to solve, like making egg-bar shoes for horses with underslung heels or squaring the toes of a horse that tends to overreach his stride.

Tim has a pickup truck in which he carries all the equipment necessary to make special shoes if he needs to. That includes a forge, an anvil, a drill press, an acetylene torch, and a generator to run the grinders.

Tim charges $27 to shoe a horse and $10 to trim the feet. He likes to be able to do six or seven horses a day but can do as many as ten.

"It's hard work," says Tim, "but you're out in the open all day, not cooped up in one place. Every day I'm somewhere different. I don't have people telling me what to do all the time. I can choose my own customers. And the money is pretty good."

Tim thinks future blacksmiths ought to start young, around fourteen or fifteen years old, working for another blacksmith. That would give the young person time to build his back muscles.

"It's really hard to learn this trade when you are over thirty," says Tim. "It's just too hard on your back."

CHAPTER XIV

Riding Instructor

Owning a horse, unless you live on a farm, is a luxury. Riding horses in shows may be one of the world's most expensive hobbies. It is not unusual for show horses to cost $3,000 to $10,000 or more. An Olympic caliber horse may be worth $100,000. There are also the added expenses of tack for the horse plus the English riding costume for the rider.

Rob Gage teaches riding in a community where many of the residents can afford luxuries, including horses. He is so well respected as an instructor that he, too, can afford some luxuries, but he does work hard for his $50,000-per-year income.

"I started show riding when I was about twelve," recalls Rob. "I worked with a good trainer and went on the show circuit. I think to learn this business you have to work with a really good instructor, one whose students are winning the competitions. It's like being an apprentice. By the time you are seventeen to nineteen, you can begin giving lessons to beginners."

Except during the winter when there are few shows, Rob works nearly seven days a week. He starts each morning by working with some of the younger horses that are in training themselves. By early afternoon, children begin to arrive for lessons. Rob teaches until 5:00 each afternoon.

"I only accept youngsters who are willing to go on the riding show circuit," says Rob. There is a younger instructor at the barn who teaches the beginning students.

During shows, Rob puts in an even longer day. Shows may last anywhere from three to fourteen days. Rob usually arrives the day before the show to make sure the horses are well taken care of and to check out the courses his students will ride. Depending upon the

Beginning riding students start with low jumps.

event, Rob may have from three to forty horses and students participating (some students have more than one horse). To care for that many horses, Rob employs grooms and assistant trainers. It is Rob's job to arrange rooms and meals for all of his employees. In addition to salary, he has to pay for workmen's compensation insurance for the grooms. With all the details to worry about, Rob is a small independent businessman.

On show mornings, Rob is up and moving by 5:00 A.M. He has to check that the grooms are preparing the right horses. He checks for any horses that may be lame or sick. He makes a final check of the course and talks with his students, giving them tips about the difficult parts of the course and how to handle any problems. He has to be at the ring area whenever his students are scheduled. Some shows last until 10:00 P.M.

By Sunday night Rob is ready to head for home, but he still has responsibilities about the care of the horses and tack. Monday morning, the day he is supposed to rest, finds him still putting away the equipment, checking the horses, and then exercising the horses that didn't go to that particular show. During the show season, this routine goes on about three out of every four weekends. Rob wouldn't have it any other way.

"The best part about the job is working with the horses and being outdoors all day," says Rob. Even though he seldom gets a day off, he likes being able to choose his own hours of work. But there are problems to be managed.

"When a parent invests a lot of money in a horse and training for his child, he really expects the child to win," explains Rob. "It's hard to tell a mother that either her child or the horse doesn't have enough ability to win. You just have to expect some clients to get unhappy with you and move on to other instructors."

A large part of the success of a riding instructor is his ability to sell himself. According to Rob, that takes both talent and charisma as well as an ability to get along well with children.

Judy Martin is a riding instructor at a barn not far from Rob's facility. She agrees that parents are likely to present a far larger problem than any of her young students. Horse shows, she thinks, bring out either the best or the worst in people.

"Obviously this sport is like any other," she says. "The point is to win. But I like to let each child perform at his own level and enjoy it from there. You can make riding just about anything you want it to be. The family makes the choice. You don't have to invest enormous amounts of money in it for it to be a good learning experience."

Judy started instructing others as a way to help support her own horses. Little by little she gained experience and a reputation as a competent instructor.

"Time has been my best instructor," she says. "I've learned through trial and error. A young person can get a feeling of independence from this sort of work. You can start by doing free-lance teaching, charging maybe $3 or $4 per lesson. That's more than you can earn working at most part-time jobs. You can build up your business, and at the same time you are able to choose your own hours of work. After you gain some experience, you can charge from $12 to $20 for a private lesson."

In order to succeed as a riding instructor, Judy says, "You have to like horses, children, and people in general. It's very rewarding to see a child who at first couldn't even walk or trot on a horse come along and do well."

Cyndi Grossman is one of those students who did well and is now an instructor herself. She started riding when she was ten. To help pay for lessons and the care of her horse, she began working around the barn for Rob and eventually taught the younger children at his facility. She attended college for two years, working part time as an instructor. Finally she decided that college would just lead her to a desk job, so she devoted herself full time to the riding business. Now she works for Judy.

"I got really good basic training in the Pony Club activities," says Cyndi. Pony Club is an international organization that encourages riding activities for youth. "And working for a good trainer like Rob gave me more opportunities to learn. I've managed to turn my hobby into a career, and that's pretty hard to beat."

Cyndi starts her mornings by working out green or untrained horses that she has purchased. After training them, she sells the horses, making a nice profit that she usually reinvests in other horses. She also handles adult students during the morning hours before the after-school crush of youngsters.

"We emphasize the whole area of horsemanship at this barn," says Cyndi. "The kids have to learn all about the care of their horse and tack as well as how to ride. It's through grooming and caring for your horse that you really get to know the animal."

One of the important things Cyndi has learned is never to put an inexperienced rider on an inexperienced horse. Beyond the fact that the safety of both horse and rider are in her hands, it is often the good horse that can best teach a young rider.

"Communication is the most difficult part of the teaching job," she says. "Sometimes it is very difficult to explain to the rider what you want him to do in a way that he can understand and comprehend. The best thing is seeing the kids and the horses I have trained progress and become really good."

Cyndi encourages young people interested in the business to read as much as they can about horsemanship and to go to as many shows as they can to learn skills. She points out that there are schools in both northern California and Virginia where horsemanship is taught, but she feels that learning from a good instructor and being on the show circuit yourself is probably the best way to gain experience. If necessary, she thinks, young people should work at a barn in trade for either the boarding expense of their horse or the cost of lessons.

Cyndi's dream, should the opportunity ever present itself, is to run her own stable.

CHAPTER XV

Pack Station Operator

The sun has just begun to touch the tops of the pine trees, but already packer John Cunningham has five mules loaded with camping gear for the four-member party he will lead for a week through the Sierra high country of California. The mules each carry 150 pounds of equipment that must be lashed into place, carefully keeping the whole load in balance. A top-heavy load can mean trouble on the trail.

By 7:00 o'clock the four adventurers, seeking fresh air, grand vistas, a few golden trout, and the quiet of the outdoors, are ready to mount their saddled horses. One man has never ridden a horse before. John provides brief instructions on the management of the animal and resolves to be patient with this beginner. Then he starts the party on the trail and rounds up the five mules that he will lead. They are balky at first, unwilling to leave the corral area. Once on the trail, however, they plod along, roped in single file, each step taken surefootedly.

John Cunningham has been leading pack trains as long as he can remember. His father bought the High Sierra Pack Station permit from the National Forest Service the year John was born, 1948. John was on the trail before he was five years old and went on trips alone with experienced clients by the time he was six.

Near the pack station, the trail is wide from the frequent use of both horses and hikers. Even so, a doe is spotted not fifty feet from the trail. It is startled by the parade of horses, men, and mules and dashes through the undergrowth out of sight.

By 11:00 o'clock horses and men are weary. They stop for lunch beside a mountain stream. As they eat, clouds build on the 9,000- to 14,000-foot peaks to the east, bringing the threat of an afternoon thundershower. Later, as they continue to their first night's camp, they meet a lone backpacker hiking the same trail. He stands beside the trail as the horses and mules pass. Just as the middle mule is opposite him, the hiker decides to put on his bright orange rain poncho. The mule spooks and tries to run away. That causes his fellow pack animals confusion and fear. Amid much braying and kicking, the lead lines are tangled and finally one mule falls, his load slipping sideways under him. It is nearly an hour before John can get the mules moving again.

That evening John unpacks and unsaddles the animals. They will graze freely on the mountain grasses until they get their fill and then sleep until dawn. The men eat their evening meal around a small campfire and exchange stories of their past adventures. John goes to his bedroll early because he will be the first one up the next morning.

The unfettered horses and mules wake early. If John isn't there to round them up, they will begin walking and grazing as they go. Packers have been known to trail their straying horses for fifteen miles, sometimes all the way back to the pack station, when they have been careless about oversleeping.

John puts the heavy loads on the mules and saddles the horses. The party is on its way again.

Tom Cunningham, now a wind-worn man in his fifties, was raised on a ranch. He began to learn the packing business with a cousin who had one of the first pack station permits in this part of the Sierras. On the western slope, Tom's station has access to thousands of acres of National Forest land situated between Yosemite National Park and Kings Canyon National Park. Hundreds of lakes dot the landscape, and it is prime deer-hunting country. Not far from the Edison Lake headquarters, the John Muir Trail winds its way along the peaks and valleys.

Though the mountains have remained unchanged, the packing business has changed during the last twenty years.

"We used to run 120 animals from this station," says Tom, "but now we have only 80. There are more backpackers now who are seeing the mountains on their own two feet. That didn't happen until new dehydrated foods were developed along with a better backpack. The trend toward more recreational and four-wheel-drive vehicles has cut into our business, too."

Neither Tom nor his son John seems too concerned that business is less now than it used to be.

"Running a pack station is a good excuse to get out of the valley during the hot summer," explains John. "I like animals and people and the mountains. This is a way to make a living up here."

Running a pack station in the Sierras is not a year-round job. Eight to twenty feet of snow cover the hills and valleys during the winter, and it is only from June to October that the back country is accessible. During the winter season, father Tom is a carpenter and John is a truck driver.

"I like being a packer in the same way I like driving a truck," says John. "Once you are on your way you are on your own. There is no one looking over your shoulder telling you what to do. It's your responsibility to get there and back."

The busiest time of the year for the pack station is the six-week deer-hunting season that starts in mid-September. And about 85 percent of the business throughout the year is "spot" trips where the packer takes his party to a site with all their equipment, leaves them there for the agreed upon time, and returns to the station with the unloaded animals. Since each animal costs the client $20 per day, it is understandable that most of the trips are "spot" rather than "continuous," with the packer and animals staying with the party the entire time.

Any trip, however, provides opportunities for a certain amount of excitement or danger.

"Probably my most potentially dangerous situation was when I was leading a party back from a trip," recalls John. "We got caught in a real downpour. The creeks were badly swollen. At one point where we were supposed to cross the creek, the footing had washed out and a log had washed into the area. The lead man tried to make the crossing and the horse's feet went out from under him. Both the rider and the horse were swept downstream. Eventually, the rider separated himself from the horse. (It is dangerous to try to swim with a horse, since he may panic or kick you.) Both of them ended up safely on the other side of the creek, but the rest of us were stuck on the wrong side until I could find a better crossing."

Another occasion that both Tom and John recall vividly was when they had been hired by the Sierra Club for a pack trip in Navajo country near the Grand Canyon in Arizona. The entire party, including Tom and John, were caught by surprise the first night out when a late spring blizzard struck. No one was prepared for cold weather, and the entire shivering trip had to be canceled.

There is a certain amount of danger right at the pack station. Broken bones—arms, legs, ribs—are not uncommon for these two men who have broken their share of horses to saddle.

The mountain season begins for the Cunninghams about June 15 each year, when they and a good many friends drive the herd of horses and mules up the mountain. The three-day trip covers 70 miles from the valley pastures to an elevation of 7,000 feet. Although six or seven riders would be adequate, it has become a traditional spring affair for all the friends of the family.

One year Tom arrived early at the pack station to discover that a bear had wintered in their cookhouse. A restless sleeper during his hibernation, the bear had destroyed two refrigerators and redistributed all the kitchen utensils. Not exactly housebroken, the bear had left a three-inch-deep mess throughout the building.

Fortunately, Tom's wife, Emma, did not arrive until several days later, after Tom had had a chance to clean up most of the mess.

Though Mrs. Cunningham seldom rides, she is a very important part of the pack station business. She keeps the records, writes letters, handles reservations, and answers written inquiries about the best hunting or fishing destinations. She is also usually the one to drive down the mountain to Fresno, the nearest large town, for supplies.

"If I were a real social person, a club joiner, the remoteness might bother me," she says. Her nearest neighbors own a small resort and boat marina about two miles away. "We live like a big family here with the cooks and the boys who help out. I rather like the quiet. It's not as if you were all alone. You meet a wide variety of people—doctors, Congressmen, Boy Scouts, Sierra Club members. They are all nice people."

The only negative impact of pack station life on the Cunningham family was while John was still in school. He had to live with friends for the first six weeks of school each fall while the family handled deer-hunting trips.

Most of the young wranglers the Cunninghams have hired over the years, who are paid very little beyond room and board, have been relatives, often John's cousins.

"Each summer we start a couple of boys who are about fourteen," says Tom. "Every year they learn a little more. They usually stay with us until they are eighteen to twenty-two. Then they decide they have to have a regular year-round job, and some of them have been hired by the Forest Service."

Recently Tom has started to hire young men who are not relatives, some the sons of regular clients.

"If someone wants to be a wrangler, he should first learn to handle stock, like on a farm," says Tom. "He should at least know how to saddle and care for a horse, which is something that can be learned in 4H programs. I also understand that Bob Tanner at Red's Meadow and Mike Knapp at Fish Camp on the eastern slope of the Sierras have summer camps where they teach kids to become packers. The kids pay tuition like at a summer school and work as swampers (general helpers) on actual pack trips."

The National Forest Service is no longer issuing pack station permits, so anyone who is interested in going into the business must find a seller of an existing permit. There would be many factors to consider.

"First of all, it depends on how you want to make your money, either on long trips into the back country or on one-hour or two-hour trips for general tourists," explains Tom. "The most important thing is that your pack station be accessible by car, and then, if you are going to emphasize the long pack trips, that the back country is easily accessible from your station."

The vast majority of pack stations are in the eleven Western states. Those in the heavily visited National Parks emphasize short trips, often back and forth over the same trails every day. Stations in the hunting areas of Wyoming are governed by state laws that allow only one commercial guiding outfit to serve a particular area. In California the pack station operators can move around, crossing paths, although there are gentlemen's agreements about the frequency of trips into competitive areas.

The services provided by packers vary, too. In California the clients generally provide their own food (and enough for the packer), sleeping bags, tents, fishing equipment, and other gear and must cook their own meals. Typical Wyoming packers provide more services and equipment, and their prices per day are presumably higher. (The Cunninghams provide full service, if requested, for an increased price.)

The Cunninghams meet expenses with their pack station efforts but will never be rich. For them, a summer in the mountains is reward enough.

PERFORMING IS THE GAME

CHAPTER XVI

Hollywood Animal Trainer

Bryan Renfro got into the animal-training business almost by accident, yet it clearly was in his blood.

Bryan's father was one of the earliest animal trainers in the movie industry, as well as a stunt man. Among other animals, he owned and trained Daisy and her puppies who were featured in the *Blondie* series.

Bryan's uncle, Ray Berwick, is still active and one of the leading trainers in the business. Specializing in birds, Ray has created some memorable scenes in movies such as *Bird Man of Alcatraz* and the Hitchcock classic *The Birds*.

"I got out of the service with a wife to support and no job," recalls Bryan. "Ray, just as a favor, gave me a job at $50 a week cleaning cages. It was supposed to be temporary until I found something else, but I really started to get into it. I'd ask for something to train and Ray would tell me how to do it."

Bryan learned by doing and learned his craft well. He is now head animal trainer at Universal Studios Tours and a regular contributor to movies and television shows, including *Little House on the Prairie* and *Barretta*.

"Not everyone can train animals," explains Bryan. "Oh, I suppose everyone can train an animal to a certain extent, but to teach the complicated things we ask of our animals, training has to be in your heart and in your mind. You have to be able to get into the mind of the animal. Every one of them is different; you have to be able to understand the animal. More than just training, you have to be able to communicate."

Bryan uses the positive method of training an animal, with food as the reward. Some discipline may be necessary when the animal is not paying attention, but basically Bryan feels that patience is the key to guiding a dog or a bird back to the desired behavior.

"Hitting or yelling at a dog really reflects the frustration of the trainer and maybe relieves some of his tension, but it doesn't teach the dog anything but hate," says Bryan. "Basically, an animal wants to please you. It is its nature."

"By their very nature, animals want to please their trainer," says Bryan Renfro of Universal Studios Tours.

Bryan and Ray specialize in birds and dogs (dogs are the most frequently used animals in the movie industry) but have also trained raccoons and rats. They do not usually train the more exotic animals such as leopards, snakes, and elephants.

"Pound for pound, birds, particularly cockatoos, macaws, ravens, and crows, are the most intelligent animals around," says Bryan.

Ray Berwick was the first trainer to teach birds to fly into the wind created by a large fan, making possible movie scenes of birds that appear to be in actual flight. Previously, birds had been held with their wings spread and pictures could be taken of only a part of the bird in an effort to simulate flying.

Training an animal for a movie or television trick may start several months before filming. Bryan is contacted by a studio and told of an up-coming need. If the trick is very specialized, he will begin training, using the moments of spare time that are available to him between shows at Universal Studios. At other times, he is told right on the set what is expected of his animal for that day's shooting.

"The easiest thing in the world is to teach an animal to go pick up something, carry it to a table or other spot, and put it down," reports Bryan. "For a *Barretta* show the director wanted the cockatoo to pick up a box of cereal and carry it to Barretta's bed. It took me ten minutes to teach the bird to do that. What is far more difficult is to make a talented, well-trained dog (like Bandit on *Little House on the Prairie*) look like just an ordinary pup. The dog keeps looking at you for commands and what you really want is for him to look and act just like the dog down the street."

According to Bryan, there are many tricks of the trade that make the animals look good on film. Usually when an animal is featured, there will be a wide shot that includes the human actors in relationship to the animal. Then there will be a close-up of the animal doing his trick. During the close-up the trainer can be near the animal giving verbal commands, and there will be few distractions. It is far more difficult to involve an animal in a wide shot that requires the trainer to be at some distance to give commands. To solve some of those problems, Bryan can teach the animal to respond to a buzzer or even to follow the natural, scripted commands of the actor. Hand signals are used, too.

"Sometimes I have to give a voice command right in the middle of the script," explains Bryan. "I try to work it out with the actors and the director so that my voice falls right between two of the actors' lines. That way they can erase my voice in the editing process."

From time to time Bryan finds that his movie career offers some difficult challenges and some danger. For an Andy Griffith movie for television, Bryan's golden eagle was to attack one of the actors by landing on the end of a rifle. Eagles do not naturally have any liking for rifles. In the same movie Bryan doubled for an actor in a scene that found him in a small plane along with a wolf and the eagle. The plane had just crashed and was burning. Bryan's task was to keep the animals calm in the face of fire while they got themselves out of the plane.

In still another movie Bryan was playing the role of a bad guy. As he pointed a hand gun, his eagle swooped down on his hand. "I wanted to do the shot without gloves on, which was a little crazy," laughs Bryan. "In the rehearsal the bird landed on my hand and his talon went in two inches deep. I knew we had to do the shot right away, because within minutes my hand would be all swollen and we'd never make it."

Even with injuries, a fairly common occurrence, Bryan likes the chance to do a little acting and stunt work, just for the variety it adds to his days. He has also appeared on several television talk shows.

Bryan's day at Universal Studios takes on a different look from his days on a movie set. There he is a showman, a performer whose stage presence and chatter entertain and educate thousands of visitors each year. Typically, he arrives at the Studios about 9:00 A.M. He and his two junior trainers clean the cages and wash the dog runs. During the winter months, they perform four shows a day. During the busy summer tourist season, they put on a frantic twelve shows each day and often do not leave the Studios until 10:00 P.M. Still, Bryan likes to work in front of an audience where he can respond to the laughter and the applause.

Both of the young men who work for Bryan attended Moorpark College near Los Angeles, which offers a special class in animal training. Although Bryan does not think the school program teaches many of the things a trainer needs to know, the course did provide these young men with at least their first contact with the industry and a way to show that they were seriously interested in a career in animal training.

"I look for someone who is willing to work hard," says Bryan about the young people he hires. "You can get lazy in this business and just do the shows. I like to see people working and learning every chance they get. Most new people start with cleaning chores and are paid very little. I usually start trainers with birds, and they had better be prepared to be bitten, particularly on the face. It takes a little more experience to train dogs."

Bryan explains that amateur animal trainers seldom get a chance to work in movies or make television commercials, although he encourages young people interested in the trade to train their own animals.

"A production company really wants to work only with professional trainers," says Bryan. "They call me, I don't call them. They know that if I say I have a dog that can do a certain trick, it is true, because I'm a professional and my reputation is built on success. If a stranger calls a studio and says he has a cute dog with a clever trick that ought to be in a commercial, the chances are good that under the lights and with all the confusion on a set the dog would not be able to perform."

Bryan thinks the best way for a young person to get started is to work for an established trainer. Lists of trainers and how to contact them are available in what is called a *Studio Directory,* which is used by casting offices and producers. An ambitious would-be trainer could get hold of that directory through stores that rent commercial movie cameras or by simply talking with someone like Bryan.

For someone who takes the initiative, works hard, and has a lot of fun training animals, the sky is the limit to the financial rewards.

CHAPTER XVII

Marine Animal Trainer

The dolphins circle anxiously in their holding tanks. It's almost show time. The gates are raised. With a wave of his arm, Jeff Zimmerman directs the three lead dolphins into their first unison leap and the show is on at Hanna Barbera's Marineland near Los Angeles, California.

Jeff has been trainer and performer with dolphins, sea lions, and killer whales for the last six years.

"There is more to this job than just training," explains Jeff. "I'm a performer, writer, director, set designer, and kennel cleaner."

Jeff's day starts early with the routine chores of feeding and vitamin-giving for the sea lions and dolphins. Only part of the daily diet is given in the morning. The rest is saved to be used as rewards for good performances during the day.

Since sea lions spend the night in dry cages, Jeff must next clean those cages, an unpleasant but necessary task for everyone who deals with animals.

During the day Jeff performs four shows, two in the morning and two in the afternoon. And the show goes on whenever there are paying customers, even in fog or cold. On rainy days, when there are no customers, Jeff still works the dolphins through one or two show sets just to keep them sharp.

In between shows Jeff works with the new animals, training behaviors that will eventually produce star performers.

When the customers leave there are more clean-up chores to be done. The pools must be drained and cleaned alternately, which means that Jeff climbs into the empty pool to brush and scrape the sides and bottom.

If there are sick animals, Jeff spends part of his day, and sometimes part of the night, helping the veterinarian take blood samples, give shots, and force-feed.

Jeff got his job almost by accident.

"I knew there was a job open [at Marineland], but I didn't really expect to get it," recalls Jeff. "I applied just to get some practice applying for jobs. I was going to school, majoring in criminology. Much to my surprise, they called me back in a few days and offered me the job. I'm a natural ham, so it was easy for me to put on a microphone and do the shows."

Jeff is perhaps too modest. He did have a good background in basic psychology, an important part of training animals, and he had unpaid experience training dogs. He was first assigned to work with dolphins, which are the easiest to train. Then he moved on to killer whales.

"You have to be wary of killer whales," says Jeff. "You know they will misbehave. You just don't know when."

Jeff makes the training of a newly captured animal sound easy, but it is a nine- to twelve-month patient process before an animal can learn six to twelve behaviors that make him show-ready.

"The first step is getting a captured animal to eat," explains Jeff. "They are used to catching their own live food. Here everything is dead, frozen. It doesn't usually take more than two or three days before the animal is hungry enough to try the dead fish. Once they try it and find it meets their needs, there isn't any more problem."

All of the training is done through conditioning and behavior modification, with food as the reward for proper behavior and no food the penalty for misbehavior.

Each time the new animal is fed, the trainer blows a whistle. That conditions the animal to associate the whistle with the pleasant experience of eating. The trainer can then use the whistle to bring the animal to him.

Then a ten-foot pole with a foot-long Styrofoam tip is introduced to the animal. This becomes the target. Each time the animal touches the tip with his nose, he is rewarded with some more delicious fish.

"Once the animal is used to the target, you can move the tip around, getting the behaviors you want out of the animal," says Jeff. "You start at water level. Then you can move the tip higher and higher until you get leaps or back flips. While you're doing that you also use hand signals. These animals are so smart, catch on so fast, that you can soon eliminate the pole and they respond just as well to hand signals."

Each of the animals has a distinct personality, just as each person is unique.

"The animals have favorite trainers," says Jeff. "They will perform at a higher level for one than the other. And sometimes they are lazy. They'll look for ways to do as little as possible to get as much as possible."

Jeff thinks of his dolphins and killer whales as "almost buddies, friends."

Actually doing the shows gets to be the most tedious part of the job.

"In training there are challenges; there are obstacles to overcome," says Jeff. "We only change the show about once a year. After you do the same show two or three times a day for three or four months, it gets old. It's hard to stay up for the show, particularly if it's a cold day with a small audience that can't clap because they've got their hands in their pockets to keep warm."

There are some natural hazards in the job, too. Jeff is in and out of 55° water several times a day, sometimes unexpectedly, like the time he overshot the dismount platform after riding two dolphins surfboard style around the pool. The platform was particularly slippery. He went sliding right on off the other side to join his dolphin friends in the water. He just had time for a quick hot shower and change of clothes before his next performance with the sea lions.

With slippery decks, strained muscles are commonplace. More serious are the accidents with killer whales. Their sharp teeth can rip away a lot of skin, requiring the trainer to have stitches.

The job also means that Jeff spends a lot of time smelling like a tuna boat, a perfume that isn't particularly appealing to some of the girls he dates.

Even though the shows may be tedious, there can also be surprises.

"You can be out there with a stadium packed full of people," laughs Jeff. "Your dolphins are all seasoned performers. Then all of a sudden the animals go to the bottom of the pool and quit. It's like they all went on strike. I'm not a very good ad libber. You can really feel like an ass out there."

After an incident like that, there isn't much Jeff can do but go backstage and laugh it off. Fortunately, the next show is likely to go well, since the animals will be hungry and ready for their fish rewards.

With all of the running noses from being chilled regularly, the strained muscles that are an occupational hazard, and the performances that don't always go just right, Jeff isn't about to change jobs.

"The surroundings are beautiful," he says as he looks out over the cliffs that fall into the Pacific Ocean. "You're outdoors 90 percent of the time. You're working with animals that have magnetism, a real mystique about them. I'm pretty much my own boss in my area. We all listen to each other's ideas when it comes to new proposals."

Jeff thinks that someday, after he has learned a lot more about the business, he would like to be director of training at some marine park. That would mean an income of about $20,000 to $25,000 per year.

For young people interested in becoming marine trainers, Jeff has several recommendations.

"They should start young by raising and training their own animals," suggests Jeff. "They can train exotic birds, dogs, horses, anything, and show that on a résumé when they apply for a job. They should also read books about training animals."

College training is not required to become an animal trainer, although trainers are becoming more professional and knowledgeable in their field. Psychology, not marine biology, is the important course to take.

"You pick up, almost automatically, what you need to know about animal physiology, on the job," he says. "Psychology courses help you understand the learning process through rewards and behavior modification."

Jeff strongly recommends a two-year program offered at Moorpark Community College near Los Angeles, California. Bill Brisby conducts the courses in care of wild and exotic animals and is able to place 70 to 75 percent of his graduates in jobs at zoos or animal parks. A similar course is offered at Santa Fe Community College in Gainesville, Florida.

"Young people ought to just hang around a marine park if they can," says Jeff. "When kids ask me for advice, I tell them to buy a season pass (it's relatively inexpensive) and become an observer in the public feeding areas. They should ask questions whenever they can and be there so much that they are almost a nuisance. When it comes time to hire someone, I'll know this person is really interested and knows something."

Beginning trainers do not earn large sums of money. Marineland's pay rates are relatively high, largely because the employees are unionized. Other parks may start their trainers as low as $500 per month.

When Jeff started in the business, there were few if any female trainers. Now there are several, including the two employed at Marineland.

"I don't know why it is, but the women seem to have a special knack with these animals," says Jeff.

BUSINESS

CHAPTER XVIII

Pet Store Owner

Morning is a busy time at the Pride and Groom pet store. The puppies and kittens, delighted to see people again after a night alone, are eager for breakfast. Their cages must be cleaned to make them fresh-smelling for customers who will soon be arriving.

Before 9:00 o'clock, grooming customers begin to arrive with lap-size, middle-size, and huge dogs for haircuts and shampoos. Some dogs are willing victims of the bathtub, but most are nervous and add to the excitement level in the store by resisting. They are wrestled into the tub with frequent splashes of water hitting the groomers and the floor.

Meanwhile, another customer arrives seeking medicine for her pet parakeet who seems to have a bad cold. Still other customers arrive to buy 25- or 50-pound bags of feed for their animals, new filters for their fish tanks, mice to feed to pet snakes, or a new kitten for a child's birthday present.

The center of attention and activity at Pride and Groom is the owner, Larry Canter. Between helping customers and directing the reluctant dogs to the grooming area, Larry does a quick inventory of his supplies, fixes a broken drying cage, checks on a sick dog, and repairs a cat-climbing tree. As he works, he plans changes in his displays, from featuring summer flea and tick supplies to a special on fall sweaters and coats for chilly dogs. He even finds time to pet the store's mascots, a dog and cat who have the run of the place.

Larry has owned the Pride and Groom for nine years and has more than doubled the size of the store in that time. Being a pet store owner was not his original career goal.

"I was a preveterinary student," recalls Larry. "It's really hard to get into vet school. Your grades have to be so high. And I decided I didn't want to cure all the sick animals in the world. I just wanted to work with animals."

Larry's friend, a pet products distributor and salesman, talked about the pet store business enthusiastically. When a relative agreed to lend Larry enough money to get started, Larry began a career that seems to fit his needs.

"I like working with both the animals and the customers," says Larry. "I like this better than being a salesman. And I like being my own boss."

Being his own boss means that Larry spends a lot of time at his work, at least six days a week plus some evening work at home when he does the bookkeeping and correspondence. (He hires an accountant to do quarterly reports and tax returns.)

"This is not like a men's clothing store where you hang a suit on a rack and just leave it there," explains Larry. "Things keep changing, and, of course, the animals have to be fed every day, even holidays."

Nothing in Larry's background, either his college studies or his part-time work for a veterinarian, has helped him very much as a pet store owner.

"You just learn as you go," says Larry. "You read labels, talk to salesmen, and listen to what your customers say. There are seminars sponsored by the industry that can help, too."

Larry shares the work load with four or five employees each day. Since he likes to be free to solve problems, help customers, and do some of the caring for the animals, he hires others to do the clean-up and grooming chores. Most of his employees are young. He finds that teenagers, some as young as thirteen, are more willing to do the necessary dirty work around the store and stay with him longer than adults. He has three young people, ages seventeen to twenty, who have been with him for more than three years.

"I don't like to do a lot of cleaning and the routine chores," explains Larry, "so I'd rather hire the work done and then check to see that it is done right."

Other pet store owners use a different approach. Many operate the store as a family affair, employing only one or two people outside the family group of husband, wife, and children.

Selecting the products the store will carry, both supplies and live

The author's daughter Patty has a part-time job in a pet store. Here she removes a guinea pig from a cage preparatory to cleaning it.

pets, is a learning process. Larry has found that salesmen approach him with new ideas.

"Last year a salesman offered goats and sheep at Easter time," recalls Larry. "We tried them and they went quite well, so we'll do it again next year."

The dogs, cats, and birds are usually supplied by local breeders. He likes to deal with the same breeders over the years so that he knows something of the history of the animals and can protect against any genetic problems that might develop later. Even with care, surprises do happen. Animals are discovered to be pregnant. Animals get sick and trips to a veterinarian are necessary. Raccoons are popular pets one year and don't sell at all the next.

Compared to other pet stores, Larry's is large and well run. His income is about $20,000 per year, and he gives himself a raise each year as the store's income continues to grow. This income level puts Larry, by his estimate, in the top 20 percent of pet store owners. He is considering purchasing a second store, which would add to his income.

"When you buy a store, location and size are the most important considerations," says Larry. "The location has to be where there is a lot of foot traffic. In a pet store, although many of the sales are to repeat customers, a lot of sales are impulse buying, someone seeing a cute puppy or kitten and being unable to resist it. And a small store doesn't make enough money to be worth the effort."

For young people interested in owning a pet store, Larry suggests that the first thing to do is get a job in a pet store and begin learning the business.

CHAPTER XIX

Pet Products Salesperson

Paul Kenner has always been a good salesman. Part of the reason is that he has a gimmick that makes his customers remember him. He does magic tricks. When he calls on customers, they are eager to see his latest trick, and certainly that puts them in the right frame of mind to order more from him than they might from another salesman.

Paul had spent a lot of time in sales work. In time he tired of some of the pressures that he found as well as the uncertainty. Since he had always been interested in animals and felt that would be a stable business, he sought out pet products distributors for possible employment. He is now sales manager of Merchant's Pet Supply, a wholesale distributor serving pet stores throughout southern California and Las Vegas.

"The ten salesmen who work for me have routes of thirty to forty stores that are their responsibility," Paul explains. "They visit each of those stores regularly, the larger ones once a week. They explain specials to the store owner, walk the shelves to see what is running low, and generally try to help the owner make money."

Merchant's Pet Supply carries a product line that includes items for hamsters, dogs, cats, birds, and fish. It includes such items as flea collars, vitamins, collars, cages, and treats.

As sales manager, Paul doesn't have any special route of stores to cover. He checks out new accounts, making sure everything gets started right. He also has regular meetings with his sales staff to explain promotional items that are coming up and to motivate his people to do well. He also must handle any complaints from customers.

"Perhaps the hardest part of the job is to educate the store owners," points out Paul. "So many of the owners are just hobbyists, not business people. They don't seem to understand or use good merchandising techniques. They don't take advantage of specials or advertise as they should. So you spend a lot of time helping them to understand how to make more money."

When Paul hires a new salesperson, he looks for someone who is honest and has a nice appearance and a pleasant personality. It helps if the person has some knowledge of the pet business, perhaps is a former employee in a pet store. A college education is not important to him. He has had both successful men and women working for him.

"I think a successful salesperson is someone who is straight with the customers," says Paul, "someone who doesn't oversell and let the customer get stuck with products he can't get rid of. You have to treat every store owner equally, whether it is just a small store or a chain of stores. And you have to be willing to give personal service. I have even been known to ring up sales on the store's cash register when the owner was busy. Success means an attitude of willingness to help."

Paul suggests that young people interested in selling pet products get the names of distributing companies from local pet stores and then apply for a sales position with those companies. Most of his sales people earn $15,000 to $20,000 per year. A few earn as high as $25,000.

PUBLIC SERVICE

CHAPTER XX

Animal Shelter Employee

A brush fire breaks out in the tinder-dry hill area north of Los Angeles. The area is rugged and difficult to reach. The fire rages toward populated areas where the lives of horses as well as people are in danger.

Animal control officers respond to the emergency much as the fire department's rescue squad does, racing to the scene with horse trailers to evacuate as many animals as possible. Nearby residents, those whose land and animals are not threatened, offer their horse trailers to assist in the safe removal of livestock.

Results: Several thousand acres burned, loss of nine structures, no loss of life or livestock.

A few months later heavy rains hit the Los Angeles area. With the ground cover burned off, the bare soil is unable to hold the moisture. Flooding is widespread. Wild animal compounds, housing animals used by the movie industry, are threatened by rising water. Animal control officers respond again, this time to move lions, tigers, and chimpanzees to higher ground.

"One of our problems is convincing the public that we are more than just dogcatchers," says Leslie Mitchell, fourteen-year employee of the county animal control department and now senior control officer in charge of one of the six county shelters. "Animal control is a much larger problem than most people will admit. In any city or county it is likely to be in the top three most frequent complaints. Everybody wants the problems solved, but nobody wants to spend the money to do it. Our staffing is too low to meet the challenge of how to limit the number of animals so that you aren't forced to eliminate any."

Les supervises the work of thirty-two employees at this shelter. About half of them are animal control officers who spend their days out in the field responding to complaints. (About 20 percent of the animal control officers in Los Angeles County are female.) Most of the other employees are kennel attendants, whose job it is to keep the kennel area clean and the animals fed and cared for. One animal health technician is employed by the shelter, plus several clerical employees.

There aren't many job openings per year at Les's center, but when he does have a chance to hire a new employee his biggest concern is to hire people who won't abuse animals.

Since the job is part of the county civil service system, each applicant, in addition to being a high school graduate, must take a written test and be interviewed. The written exam tests general knowledge and reading ability.

"There are so many reports necessary these days," says Les, "that it is really important for employees to have a good command of the English language."

The interviewers look for people who talk easily since the officers face many delicate public contact situations. And animal control officers are frequently asked to speak to clubs and youth groups.

Once selected, new animal control officers and kennel attendants receive three weeks of training. During the first two weeks they learn about the law and animals, how to identify breeds, recognition of sickness and disease, and how to handle the public. Then they receive a week of animal handling experience before assignment to a particular shelter. Les estimates that it takes an additional two months of on-the-job experience for an animal control officer to become minimally competent.

Every morning, each animal control officer, in his specially designed pickup truck, leaves the shelter with a list of ten or more service requests to be filled. It might be a dog running loose, a dead animal to be picked up, a cat up a tree, an alligator in the sewer, a bull with a leg stuck in a cattle guard, or cattle running free on a highway after a cattle truck has jackknifed, turning the whole rig over.

"Sometimes the public misunderstands what we are trying to do," explains Les. "For instance, we had a report of a child's being bitten. The officer spotted the dog but had trouble catching it. Finally he chased the dog for miles in the truck until the dog was exhausted. A woman called to complain about such inhumane treatment of the dog.

We had to explain about the bitten child and how important it was to catch that dog. Then she understood better about our problem."

Throughout the day the officer receives many more requests for service over his two-way radio.

There are several ways to catch an animal.

The primary catching tool is a rope lasso, which the officer flicks over the head of the animal (not cowboy style).

"Most dogs will come to you if you don't act aggressive and if you get down to the animal's level by kneeling or squatting," explains Les. "They begin to accept the reassuring sound of your voice. Once you can pick them up and get their feet off the ground, you're pretty much in control, as long as you control the head."

Virtually all animal control officers have bite and scratch scars suggesting that the catching of animals is not as easy as Les makes it sound.

"Every time I've been bitten it has been my fault," admits Les. "You have to remember that the animal is scared and trying to protect itself. You just can't get complacent."

After the officer catches the animal, he tries to find the owner.

"Our job is to get the animal back to its owner, not bring it to the shelter if we can help it," explains Les. The officer talks with the owner, explaining the problem and the law regarding stray animals, and may give the owner a citation (like a traffic ticket).

Most owners are cooperative, or at least polite. Rarely does an officer run across the problem one young man had to face. A distraught owner drew a weapon and threatened him. In this case, the officer beat a quick retreat to his truck and called for assistance on his two-way radio. The police arrived to talk the now barricaded owner out of the house before he could harm himself or anyone else.

If the rope-lasso technique doesn't work, an officer can set a trap to catch a bothersome stray dog or woodsy skunk. Sometimes the trap works fine, but the people do funny things.

"We had one woman calling us for six solid months about a dog running loose in her neighborhood," recalls Les, shaking his head. "We set a trap for him, and finally she called to say that the dog was in the trap. But before the officer could get there the next day, her conscience began to bother her and she let the animal go. What was really funny was that within three days she was back on the phone complaining about the same dog."

To catch wild animals or potentially dangerous ones, a tranquilizer dart, or "stun gun" as some officers call it, is available. There is, unfortunately, some danger to the animal when using tranquilizers, since the exact amount that should be injected is hard to determine.

Animal control officers are not ashamed to get help from experts when they need it. Les tells of the time they had to go into an abandoned house that was full of snakes, including many poisonous varieties. The owner had been taken to a mental hospital, leaving his hobby snakes with no one to care for them. Reptile experts from the Los Angeles Zoo came to assist in the removal of the snakes.

When the shelter ends up with animals that cannot be placed in good private homes, an effort is made to find some sort of a happy ending rather than destroy the animal. Those snakes, for instance, were given to zoos across the country. Abandoned parakeets have been donated to the patients in long-term care at a county hospital. And homeless cows have been given to the county probation farm where young people can learn about the care of animals.

Officer Marvin Stitts, who has worked for Les for eight years, has seen the worst of what people can do to their animals. Marvin is the humane investigator for the shelter, responding to reports of animals being mistreated.

"We serve one area where it is a status symbol to own a horse," reports Marvin. "Kids twelve, thirteen, fourteen are very involved in their horses, but by the time they are fifteen or sixteen, they are interested in other things. The parents think the children are still caring for the horse; they just never go out to the stable to check. The neighbors are reluctant to report poor treatment, so the situation is really sad by the time they call us."

Marvin has pictures that could be used in court if necessary of a horse that was 300 pounds underweight when reported to the shelter. Each rib is clearly visible beneath the tatty brown coat.

"The poor animal never had a chance," says Marvin, "yet usually we find out that the problem is just a misunderstanding between parents and their children." Usually these problems are quickly resolved by improved care or the sale of the horse to a more interested family.

"Our problem in the field is not with the animals but with the owners," says Marvin.

While the animal control officers are out in the field, Judy Briskman remains at the kennel to perform her job as animal health technician.

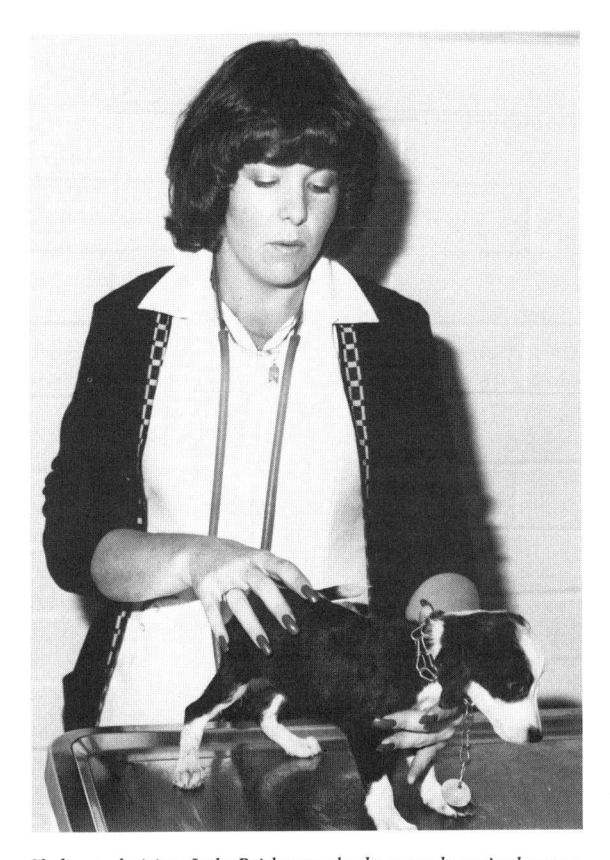

Shelter technician Judy Briskman checks a newly arrived puppy.

As animals are brought in, Judy checks each one and vaccinates it to prevent disease. If the animal is injured or ill, she provides the necessary medical care under the general supervision of a veterinarian who visits the kennel briefly each day.

Judy received her training on-the-job at a major New York animal center. If she were starting her career now, she would probably take the two-year animal health technician course at a community college.

"I love animals, I always have," says Judy. "I like to be able to help them. I've never had any fear of animals. Fear is sensed by animals and would hold a person back in this field. I think animal health technician is a good field to get involved in, particularly as an alternative to becoming a veterinarian."

After Judy checks the incoming dog, she places it in an appropriate cage. Most healthy dogs go into the large kennel building. The separate heavy wire cages allow the dogs to be either inside the building or out in the sun. There is plenty of room to move around.

Dogs that have bitten someone, injured dogs, and bitches with new litters are placed in a separate observation room.

Very young puppies and both young and old cats are placed in a third area, away from the racket of the older dogs.

Owners who have lost animals visit the shelter in a steady stream all day looking for their missing pets. Others, usually parents with young children, arrive to select a new pet. When a kitten or puppy is sold by this shelter, the new owner pays in advance for the surgery necessary to have the animal spayed or neutered six months later. "That way these animals won't be causing the birth of other unwanted animals," explains Judy.

When all else fails, when no home can be found for an animal, Judy has to put the animal to sleep. She uses the injection method.

"Ideally, there would be homes for all of these animals, but that is just not the case," Judy explains with concern. "At least if I do it, I know it's going to be done right, painlessly. I talk to them, calm them, pat them on the head. At the very least, it's better than having these animals running loose where they might be run over or become diseased."

Judy is also called upon to put very sick or injured animals to sleep. "It's almost a good feeling when I'm able to stop their suffering."

Those who work at an animal shelter are nothing like those cartoon dogcatchers, throwing a net over their hapless victims. These people are all serious in their concern for animals, their care and well-being. A sign on the side of the animal pickup truck expresses this concern. It shows a cute white and black spotted puppy behind the bars of the mobile cage. The reminder on the sign says, "One out of four of our passengers is a traffic fatality. Use a leash. We care."

THE ARTS

CHAPTER XXI

Artist / Photographer

Brushes, a palette, and oil paints are Theodora Nelligan's tools. Her product is a portrait of a family pet—a stately Dalmatian, an alert collie, a frisky poodle. Although Theodora is capable of painting almost anything, her specialty is animals.

"Frankly, I wanted to be a veterinarian, but I was just terrible in biology," says Theodora. "I do think I have a special feel for animals, though, and through my painting I can express that feeling."

Theodora particularly enjoys her work when she is commissioned to paint someone's pet.

"Usually the family decides to have a portrait done when the dog is fairly old, when he may die soon and they want a special way to remember their friend," explains Theodora. "I visit the pet in his own home as many as fifteen times, just to get a feel for the personality of the animal. Then I try to paint a more youthful animal, a reflection of the pet the family remembers as lively and playful."

Earlier in Theodora's career, she was employed by the American Needlework Institute. There she specialized in designing animal patterns for needlepoint kits, very popular items for hobbyists.

Other people with an artistic flair specialize in taking photographs of pets. They may set up their cameras in the animal's home or may ask that the animal be brought to the studio. Owners who want their dogs to be champions often commission photographs that are placed in show dog magazines as a form of advertising, hoping that judges will see the picture and look favorably upon their prize dog at the next show.

Obviously, taking pictures of animals creates some problems. Dogs and cats do not often pose well. They don't like the bright lights, and a flash may startle them. Certainly animals can be unpredictable.

Most families are content with a scrapbook snapshot of the family pet. Usually it is only rather wealthy families that will pay someone to paint or photograph their pet. That means that most artists and photographers will not be able to support themselves well by working only with animals. Theodora paints other scenes—woods, oceans, and mountains—that sell to a wider audience of buyers. A photographer will have to continue doing weddings, family portraits, or commercial work if he is going to have enough income to support himself and his family.

On the other hand, according to Theodora, there is no greater pleasure than using your artistic skills to recreate the memory of a beloved pet.